SDGs and Textiles

Editor-in-Chief

Hafeezullah Memon, College of Textile Science and Engineering, International Institute of Silk, Zhejiang Sci-Tech University, Hangzhou, China

The book series "SDGs and Textiles" addresses the strategies to achieve sustainable development goals (SDGs) in the present, past, and future. It presents books about the present and future policies of textile ministries of different countries, and books related to sustainability education around different parts of the world in the textile sector. Moreover, it would welcome the conference proceeding related to SDGs and Textiles. The series would cover books comparing the sustainability and SDGs of different institutions and countries. The individual book volumes in the series are thematic. The goal of each book is to give readers a comprehensive overview of a different area of sustainability in the textile sector. As a collection, the series provides valuable resources to a broad audience in academia, the research community, industry, and anyone looking to expand their knowledge of SDGs and Textiles.

Textiles and life are together – life cannot be separated from textiles as it is the most important need for human beings after food. In 2015, the United Nations General Assembly proposed 17 interlinked global goals to be achieved by 2030. Since then, academia and industry have paid much attention to achieving these goals. Textile found its close relation with almost all of these 17 goals.

SDG 1 - No Poverty: Poverty would never be overcome by a charity only; it is essential to develop people's skills to have a better and wealthy life. Thus, the textile can be considered an excellent discipline to achieve this goal by creating jobs and small and medium businesses.

SDG 2 - Zero Hunger: Through the effective utilization of advanced application of Agrotech Textiles, it is possible to have higher crop yields and save crops from rough weather, unexpected rains, floods, insects, etc.; thus, geotextiles play an essential in achieving this goal of sustainable development.

SDG 3 - Good Health & well-being: There has been much health consciousness after Covid19, and medical textiles assist in getting good health and well-being.

SDG 4 - Learning & Education: Textile or fashion has remained a significant discipline for societies for ages, and there has always remained much to explore in this field. Textile-related universities may play a vital role by offering free access to their education resources, training and spreading information among the locals.

SDG 5 - Gender Equality: The textile sector is one of the industrial sectors that accepted gender equality long ago; in particular, the garment sector has more females than males. Thus, the textile sector has been doing gender equality. Moreover, there has been a recent trend for Gender Neutral Clothing, which need worth studying and may further assist gender equality.

SDG 6 - Clean Water & Sanitation: Textiles could be achieved through filtration, and of course, textile is one of the critical materials for filtration.

SDG 7 - Affordable & Clean Energy: With the recent advancement in material science and engineering, the textile sector has come on the front for, not only by using this clean energy during textile production but also by assisting the production of this clean energy, either in the form of wind turbines blades made of textile composites or by energy harvesting from T-Shirts, etc.

SDG 8 - Decent Work: Recently, there has been much attention that the textile workers are not paid well, labor rights are not cared about, etc.

SDG 9 - Industry and innovation: Textile Industry always follows innovation; the textile companies that do not chase innovation cannot survive in the market.

SDG 10 - Reduced Inequalities: Getting better life and well-being would help reduce inequalities in the textile industry.

SDG 11 - Sustainable Cities: Sustainable Textile Cities through Buildtech and transport textiles.

SDG 12 - Consumption and Production: Textile and garment consumption and production all come under.

SDG 13 - Climate Action: Oekotech or Ecotech Textile, waste management of textiles are upfront to achieve this goal of sustainable development.

SDG 14 - Life Below Water: Mitigating microfiber waste in rivers and oceans may come under the context of it. There has been much attention on this subject after passing the bill at the parliament level of the UK.

SDG 15 - Life on Land: Geotech or Geotextiles studies life on land.

SDG 16 - Peace, Justice, and Strong Institutions: Protective textiles are doing their best to achieve peace, justice, and strong institutions.

SDG 17 - Partnerships for the Goals: The application of textiles to achieve sustainable development goals is only an example. In all textiles sectors, combined efforts of all the goals are essential to achieve true sustainability.

Sadhna · Rajesh Kumar · S. Greeshma
Editors

Climate Action Through Eco-Friendly Textiles

Springer

Editors
Sadhna
Department of Fashion Design
School of Arts and Design
Woxsen University
Hyderabad, Telangana, India

Rajesh Kumar
Department of Fashion Design
School of Arts and Design
Woxsen University
Hyderabad, Telangana, India

S. Greeshma
Department of Fashion Design
School of Arts and Design
Woxsen University
Hyderabad, Telangana, India

ISSN 2948-1236 ISSN 2948-1244 (electronic)
SDGs and Textiles
ISBN 978-981-99-9855-5 ISBN 978-981-99-9856-2 (eBook)
https://doi.org/10.1007/978-981-99-9856-2

This Springer imprint is published by the registered company Springer Nature Singapore Pte Ltd.
The registered company address is: 152 Beach Road, #21-01/04 Gateway East, Singapore 189721, Singapore

Preface

In a world facing an ever-escalating climate crisis, our choices as individuals and as a society have never been more critical. Climate change presents us with a formidable challenge that demands our immediate attention, innovation, and action. Our future and that of generations to come depend on our ability to make informed decisions and embrace sustainable solutions. In this context, *Climate Action Through Eco-Friendly Textiles* emerges as a beacon of hope and a source of inspiration.

This book offers a comprehensive exploration of the textile industry's profound impact on our environment and, more importantly, presents a roadmap to mitigate this impact through eco-friendly textiles. With compelling insights and practical guidance, it equips us with the knowledge needed to address the environmental consequences of our choices in fashion and textiles, offering a pathway toward a more sustainable and resilient world.

The authors of this book bring a wealth of knowledge, experience, and passion to the forefront, and their collective dedication to addressing the environmental challenges posed by the textile industry is evident on every page. They provide a well-researched and balanced view of the industry's current state, highlighting the environmental and social issues that often go unnoticed. Their commitment to eco-friendly solutions shines through as they navigate the intricate web of eco-conscious textiles, sustainable sourcing, responsible manufacturing, and ethical consumption.

Climate Action Through Eco-Friendly Textiles is a valuable resource for anyone who wishes to make a difference, whether you are a consumer seeking more sustainable choices, a textile professional eager to adopt eco-friendly practices, or an entrepreneur looking to positively impact the industry. The book is not only a guide but also a call to action, an invitation to become a part of the solution.

As we face the consequences of our actions on a planet in peril, we must recognize that the textile industry's transformation is not just a desirable goal but a necessity. It is a journey that begins with awareness and education, and this book is a vital step in that journey. It empowers us with the knowledge and tools to align our choices with the broader goals of climate action, sustainability, and ethical responsibility. We are fortunate to live in a time when individuals and communities worldwide are awakening to the urgent need for change. *Climate Action Through Eco-Friendly*

Textiles exemplifies this spirit of collective responsibility, offering practical guidance on how we can translate our concern for the environment into meaningful action. It is a testament to our capacity for transformation, innovation, and collaboration and an essential contribution to a more sustainable and harmonious future.

In closing, I extend my gratitude to the co-editors and all the chapter contributors for their dedication to this important cause and their work in producing this insightful and invaluable resource. May this book serve as a catalyst for change and a source of inspiration for all who read it. Together, we can make a difference and drive real climate action through eco-friendly textiles.

Hyderabad, India Sadhna, B.Sc., M.Sc., Ph.D.

Acknowledgements

I am deeply grateful to all the chapter authors and Woxsen University, whose support and contributions made the creation of this edited book, *Climate Action Through Eco-Friendly Textiles*, possible. My sincere appreciation to my co-editors (Rajesh Kumar and S. Greeshma) for their unwavering encouragement and understanding throughout this journey. Your support has been a constant source of inspiration. I extend my heartfelt thanks to the team at Springer, whose dedication and expertise have been instrumental in bringing this project to fruition. Their commitment to quality and sustainability aligns perfectly with the theme of this book. My profound gratitude goes out to the researchers, experts, and activists in the field of eco-friendly textiles and climate action. Your pioneering work has laid the foundation for this edited book, and your insights have been invaluable. I would also like to acknowledge the reviewers, who provided valuable feedback and helped refine the content. Your expertise has greatly improved the overall quality of this book. Lastly, I want to thank all the readers who share a passion for sustainable living and environmental conservation. Your interest in this book is a testament to the growing global movement towards climate action.

This edited book is the result of collective effort and a shared commitment to creating a more sustainable future. It is my hope that *Climate Action Through Eco-Friendly Textiles* will inspire positive change and contribute to the ongoing conversation about the critical intersection of textile production and environmental responsibility.

Thank you all for being a part of this important endeavour.

Sadhna

Introduction

Climate action has never been more urgent in a world facing unprecedented environmental challenges. The impact of climate change touches every corner of our planet, affecting ecosystems, economies, and the well-being of present and future generations. As we stand at the crossroads of environmental crisis, we must rethink and reform the very fabrics of our lives—both metaphorically and literally. *Climate Action Through Eco-Friendly Textiles* is a book born out of a collective vision to address one of the most impactful yet often overlooked contributors to environmental degradation: the textile industry. Our clothes, a symbol of self-expression, culture, and comfort, carry a hidden ecological cost. The traditional textile and fashion industry is notorious for its significant carbon footprint, water consumption, chemical pollution, and social inequalities. This book embarks on a journey to uncover the interwoven threads of textile production and environmental sustainability. It illuminates the crucial role that eco-friendly textiles play in mitigating climate change, fostering sustainable development, and safeguarding the planet's fragile ecosystems.

It is a rallying cry to inspire individuals, businesses, policymakers, and innovators to reimagine how we design, produce, and consume textiles. Chapter "Introduction to Climate Action, Waste Management, and Eco-textiles" provides an overview of the current global state and the need for waste management. Focuses on the impact of textiles in the global environment and the need for eco textiles. Discusses the importance of waste management in the textile industry, as it is one of the largest polluting industries in the world. Chapter "Impact of the Textile Industry on Global Climate Change" discusses the various impacts of the textile industry on the environment in terms of health, and safety of humans and animals, toxicity and animal husbandry, carbon footprints, social justice, and water treatment. Focuses on various stages of textile production and its impacts on society. Chapter "Eco-textiles—An Overview" provides an overview of the need and importance of eco or oekotech textiles. Impact of eco textiles in terms of the global economy. Chapter "Role of New-Generation Textile Fibres in Reducing the Environmental Impact of Textiles" discusses the various conventional natural fibers and new-generation natural fibres and their properties. Influence of natural fibres in conserving the environmental impacts. Chapter "Role of Chemicals in Textile Processing and Its Alternatives" provides an overview of the adverse effects on the environment. Discusses various pre-preparatory processes such as de-sizing, scouring, bleaching, and mercerizing, and provides insights into eco-friendly alternatives such as using bioproducts and enzymes to minimize the impacts of chemicals. Chapter "Water Consumption and Microfibers: The Biggest Threat" provides insights into the excessive consumption of water bodies by the textile manufacturing industries and the pollution caused. Also, provides an overview of the release of microfibres into the water bodies causing pollution and threatening aquatic life. Chapter "Eco-friendly Dyeing Approach: Natural Dyeing—A Need of the Hour" provides an overview of industrial dyeing procedures and their adverse environmental effects, the importance of natural dyeing, and its need in the current global scenario. Chapter "Optimized Resource Consumption" discusses the importance of optimizing resource consumption in textiles regarding minimalism and capsule clothing to avoid excessive depletion of natural resources. Chapter "Importance of Market Segmentation and Application of Oekotech" provides an overview of global market segmentation and the need, importance, and applications of eco-textiles based on segmentations. Chapter "Implementation and Environment Protection Through ISWM (Integrated Solid Waste Management) in Textiles" discusses the implementation and importance of the 3rs (reuse, reduce, and recycle). Also, provides insights into protecting the environment in terms of textile waste management. Chapter "Role of Governments and Consumers" discusses the rules and standards by the government for manufacturers and consumers to possess an eco-friendly environment, standards and certifications proposed by the government to textile manufacturers. Chapter "Eco-Textiles: The Present and the Future" provides an overview of comparing the current environmental scenario with the past and possible technological innovations to have a better waste management system and a sustainable environment. Chapter "The Conclusion: Overview of the Importance of Eco Textiles and the Commitment to Be Made to Have a Sustainable Environment" testifies the power of innovation and conscientious action. It showcases the remarkable initiatives and technologies reshaping the textile industry into a force for positive change. These innovative pathways lead us towards a more sustainable future, from recycled fibres and organic cotton to sustainable dyeing methods and circular fashion models. This book is not just a call to action; it is a guide, a resource, and a source of inspiration. It provides insights into how individuals can make informed choices when selecting clothing, how businesses can transform their supply chains, and how governments can support policies that incentivize eco-friendly textile practices.

Moreover, it demonstrates that the pursuit of sustainability is not a hindrance to progress but a catalyst for innovation and growth. The authors, researchers, and experts who have contributed to this edited book share a joint commitment to a future where fashion and textiles do not come at the cost of our planet but rather enrich it. Their insights, experiences, and dedication drive this comprehensive exploration of eco-friendly textiles and their profound impact on climate action. As you delve into the *Climate Action Through Eco-Friendly Textiles* chapters, I invite you to open your heart and mind to the awaiting possibilities. The path to a sustainable future is woven into the choices we make, the knowledge we acquire, and the actions we take. Together, we can redefine the narrative of fashion and textiles, stitching a more sustainable, equitable, and beautiful world for all. Let this book serve as your guide and companion on this transformative journey as we seek to tread lightly on our planet while making a profound impact through eco-friendly textiles.

Thank you for our expert knowledge, deep analysis, and wise judgment. As the editors of this book, our work is to integrate these research results and present these wisdom crystals to the readers in the professional field to maximize the sharing and utilization of knowledge value.

Hyderabad, India

Sadhna
Rajesh Kumar
S. Greeshma

Contents

About the Editors

Sadhna received her Master of Science in Clothing and Textiles from Banasthali Vidyapith, Tonk, Rajasthan, India. Dr. Sadhna focuses on natural fibres research and woven fabrics, dyeing, and finishing. Her recent research interests include the tactile properties of seams by evaluating the characteristics of seams, such as seam compression, seam thickness, seam bending behaviour, and surface friction of seams. In 2019, she received her Doctor of Philosophy in Clothing and Textiles from Banasthali Vidyapith, Tonk, Rajasthan, India. Before joining Woxsen University, she was associated with Swami Vivakanand Subharti University, Meerut, India, as Head of the Department and Assistant Professor in the Fashion and Textile Department. She is engaged in teaching and research of Apparel and Textile Science in the Department of Fashion Design at Woxsen University. She is actively involved in her additional post-Margaret L. Bishop Professor of Strategic Design and Management. She has been awarded the title "Excellent facilitator for students and Administrative skills" by Swami Vivakanand Subharti University, Meerut, India, and Best Researcher award ITSR Foundation Award-2020, organized by the Institute of Technical and Scientific Research.

Rajesh Kumar is a graduate specialized in Knitwear from NIFT, Mumbai, and holds a degree in Masters of Fashion Management from Manipal Academy of Higher Education, Manipal, Karnataka. His research interest is in natural fibres and new product development. He has 12 years of industry and academic experience. He has served as an Assistant Professor at Manipal Academy of Higher Education, where he worked for 5 years, and he was an Academic Coordinator at Swami Vivekanand Subharti University, Meerut, India. He works as Program Director, Department of Fashion Design at Woxsen University, Hyderabad, Telangana. He has recently published two research papers titled "Traditional Craft Skills in India—Reflection of Sikki Craft of Bihar and Pathways to Sustainability" and "Impact of Stitch Density on Tactile Comfort Properties of Seams", respectively. He has been awarded the title "Best Performer for Students and Administrative Skills" by Woxsen University, Hyderabad, Telangana, India.

S. Greeshma is a Ph.D. Scholar from the Department of Fashion Design, School of Arts and Design, Woxsen University, Telangana. She holds a bachelor's degree in Costume Design and Fashion from Hindusthan College of Arts and Science and a master's degree in Textiles from Avinashilingam University. She also has an experience of 2.5 years working as an Assistant Professor at Rathinam College of Arts and Science and has published various articles including "Eco-Design: Focal Point of Sustainable Textiles", "Minimalism—Fashion Trend for 2020", and "Sustainable Textiles".

Introduction to Climate Action, Waste Management, and Eco-textiles

Sadhna, S. Greeshma, and Rajesh Kumar

Abstract Growing global and consumer demands are posing a critical challenge to the textile industry's environmental impact. The textile industry encounters a critical challenge in the form of environmental impact due to its significant contributions to climate change caused by extensive water and chemical usage, energy consumption, and waste generation. The textile industry marks itself in third position as the largest global polluter following energy and agriculture. Its environmental footprint includes contaminations, pollution, resource depletion, and microplastics. The inclusion of fast fashion holds disposable clothing and constant style changes, which accelerates waste generation leading to landfills. The industry's other pressing issue is the high energy consumption and greenhouse gas emissions. To address these issues measures such as wastewater treatment, chemical management, water recycling, effluent treatments, transitioning to cleaner energy source enhance energy efficiency. Additionally, the eco-friendly textiles produced offer a path to reduced environmental impacts. Microplastics an another significant risk posed by the usage of synthetics affecting the aquatic systems. Mitigation strategies should focus on reducing synthetic fibre usage, thus promoting and encouraging the usage of sustainable products and production methods. Fast fashion culture leading to waste generation adds to the landfills and emission of greenhouse gases. To withstand all these, the stakeholders have to promote circular economy principles, reduce overproduction, responsible consumption, effective waste management practices, following government policies and regulations, which will lead to the reduction of textile footprints. This paper highlights the multifaceted environmental issues caused by the textile industry and how to mitigate its impact.

Sadhna
Assistant Professor, Department of Fashion Design, School of Arts and Design, Woxsen University, Hyderabad, Telangana 502345, India

S. Greeshma (✉)
Ph.D Scholar, Department of Fashion Design, School of Arts and Design, Woxsen University, Hyderabad, Telangana 502345, India
e-mail: greeshma.s_phd.2023@woxsen.edu.in

R. Kumar
Program Director/Assistant Professor, Department of Fashion Design, School of Arts and Design, Woxsen University, Hyderabad, Telangana 502345, India

Sadhna et al. (eds.), *Climate Action Through Eco-Friendly Textiles*, SDGs and Textiles,
https://doi.org/10.1007/978-981-99-9856-2_1

Keywords Climate action · Sustainable · Eco-textiles · Waste management · Pollutions

1 Introduction

The worldwide demand for textile products has been accelerating regularly with the increase within the populace and human want. The 13th Sustainable Development Goal of climate action focuses on proscribing global warming, and numerous emissions due to the industries which result in climate trade. This has been pronounced as an alarming issue in this present-day decade. The primary goal of climate action is to promote international sustainability and to make the place a better place for future generations. This is a reflection of the increase in climatic adjustments which not only affects global conditions but also affects human health and the worldwide environment [1].

One of the important causes of global climate change is the textile industry, the third largest polluting quarter globally, next to the energy and agriculture industries. The textile industry requires large amounts of water and chemical usage for its production process which leads to significant waste generation in the form of liquid, gaseous, and solid wastes [2].

The textile enterprise has served as a chief contributor to environmental degradation. The environmental impacts include groundwater contamination, soil contamination, resource depletion, air pollution, and carbon emissions. The fabric and style industry debts for a huge amount of around 1.2 billion tonnes of CO_2 emissions which bills for 10% of greenhouse fuel emissions globally. The maximum of the ecological impacts is due to the carbon emissions produced by way of the extensive textile productions in apparel and footwear industries and delivery chains generating around 8–10%, which is better than the opposite shipping and aviation industries [3].

By 2023, it is estimated that only a small amount out of 92 million tons of textiles manufactured every year is recycled and the remaining end up in landfills increasing greenhouse emissions and environmental degradations. These are under the influence of fast fashion, which cultivates a "use and throw" tendency towards inexpensive and disposable clothing. The constant change in fashion and pressure to go along led to the introduction and usage of shorter-lifespan clothing items which resulted in high turnover and as well as more wastage [4].

The various steps involved in the textile production process contribute to various impacts and pollution as depicted in Fig. 1. From the stage of raw material till the end life of the product, the process involves creating water pollution, chemical pollution, land pollution, landfills, microplastics, and greenhouse gas emissions. The sustainability issues and environmental impacts caused by the textile industry are discussed in Table 1.

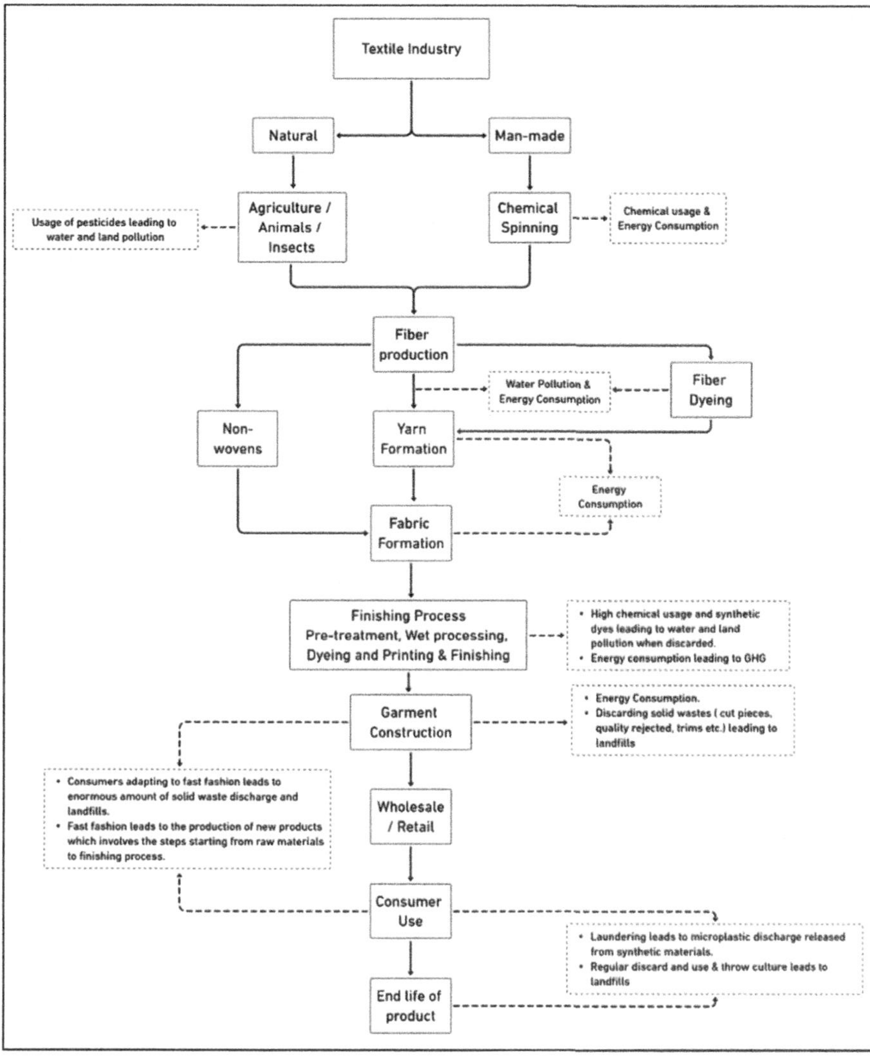

Fig. 1 Flowchart—textile production process and its impacts

2 Water Pollution—The Biggest Threat

The use of water sources in the textile industry's production process is another significant impact. It is likewise to be viewed that 20% of wastewater from the textile industry is brought about by the dyeing and finishing process. It is noted that the laundry process that is carried out for textiles produces around 51 trillion microplastics in the sea harming aquatic life [5].

Table 1 Environmental issues and impacts caused by the textile industry

Environmental issue	Issues caused	Impacts caused
Water consumption	The production and finishing process of fibres and fabrics requires a high amount of water which results in water scarcity	**Water scarcity**—depletion of water resources such as lakes, rivers, and ponds including groundwater due to high demand in the textile industry **Impact on ecosystems**—this excessive use of water results in source depletion, which in turn affects the aquatic ecosystems and wildlife
Water pollution and chemical pollution	The usage of harmful chemicals during the production process and improper disposal of the dye effluents cause water and land pollution	**Dyeing and printing**—the usage of various synthetic dyes and improper disposal leads to chemical as well as water pollution. Improper disposal without proper treatment of the effluent causes contaminations in aquatic life and affects drinking water quality for the downstream families **Wet processing**—the chemicals used for various pre-treatment processes such as hydroxides and chlorides also lead to the contamination of water resources
Energy consumption and greenhouse gas emissions	Co_2 emission is increased due to energy-intensive production and transportation and worsening climate change	**Production process**—the textile industry demands high energy consumption for manufacturing processes such as spinning, weaving, and finishing processes adding to the carbon emissions and greenhouse gasses causing climate change **Transportation**—the transportation of raw materials and product supplies also causes emissions and pollution
Pesticide and herbicide use	Crop farming relies on pesticides and herbicides, which pose risks to the ecosystem	A huge amount of pesticides, herbicides and other chemicals are used to protect the crops which cause harm to the land, bird, aquatic life and the health of farmers and nearby communities
Waste generation, fast fashion and consumer waste	The amount of waste generated by the textile industry is higher during the production and disposal of clothes	**Textile production**—the textile industry produces enormous amounts of waste such as cutting and sewing during the textile manufacturing process **Fast fashion**—the culture of fast fashion and changing trends made people buy more garments than they need. This led to an increase in textile production consuming more resources and the throw-away behavior ended up in huge landfills creating significant environmental pollution
Microplastics	The release of microplastics during laundering leads to aquatic pollution creating a threat to aquatic life and entering the food chain	**Synthetics**—fibres tend to shed microplastics when laundered and find their way into water bodies, which harm marine life and also have chances to enter the food chain risking human health as well

Water pollution is one of the stressful environmental issues the textile industry creates due to extensive water usage. The discharge of various pollutants, harsh chemicals during various wet processing procedures, and inadequate treatment methods lead to the pollution of various water bodies. This pollution has adverse impacts which affect biodiversity, habitat degradation, and even species extinction. These also affect the overall health of the ecosystem by discharging pollutants into the water bodies. Apart from the water bodies, it also poses risks to human health, especially to the communities that live near the textile manufacturing units, and can result in gastric problems, skin irritations, respiratory problems, and some long-term illnesses. This high-water consumption leads to depleting resources and creates water scarcity for the communities relying on them. Improper disposal of wastewater leads to soil contamination which in turn affects irrigation and inhibits plant growth leading to bioaccumulation of contaminants in food crops.

The textile industry requires various measures to mitigate water pollution like implementing textile wastewater treatment within the units to remove pollutants and to minimize the release of the toxic chemicals before it is discharged into the environment. Proper chemical management, proper storage, handling, and disposal prevent the disposal of chemicals into the water bodies. The industries must also ensure to comply with the water quality standards. Various practices like promoting water recycling and reusing can help reduce water consumption and overall environmental impact and reduce the demand for fresh water.

The sustainable sourcing and usage of natural resources such as natural and organic fibres requiring fewer chemical inputs helps in reducing the environmental impact and mitigating water pollution. These impacts can be reduced by making various changes in the production and consumption level and also with government intervention in textile regulations. Adopting a greener and eco-friendly approach such as substituting synthetic fibres and dyes with natural-based resources and dyes helps in reducing pollution in a greater ratio created by the textile industry. Implementing and enforcing stringent environmental regulations with proper water discharge standards, effluent treatments, and proper handling and disposal of chemicals throughout the process.

3 Greenhouse Gas Emissions and Energy Consumption: Navigating a Crisis

Energy consumption and its resulting greenhouse gases have become one of the major topics of discussions on climate change and environmental sustainability. The dependence on the primary source of energy, which is fossil fuel, leads to an alarming increase in greenhouse gases contributing to global warming and its consequences. The textile industry being a significant contributor, produces around 1.2 billion tons of carbon monoxide and is anticipated to rise by 50% by the year 2030 [6].

Energy consumption is one of the significant issues faced by the textile industry contributing to greenhouse gases and climate change. In terms of total energy usage, 38% is used by the processing industry, 23% is used by the weaving industry, 34% is used by the spinning industry and 5% is considered for miscellaneous purposes [7]. The textile industry is considered to have the lowest efficiency in energy utilization and consumes around 3/4th of the yarn and cloth manufacturing among the total electrical power required for the total textile mills [8].

4 Impact of Textile Microplastics: Unseen Consequences

In recent years, plastic pollution has become a significant debate in environmental sustainability and gathered attention to the plastic waste in oceans and ecosystems, whereas, the lesser known and equally pressing issue raised in the name of textile microplastics. These are the tiny particles invisible to the naked eye shredded from textile materials finding their way to the ecosystems as primary (< 5 mm) and secondary sources and causing risks to aquatic life and ecosystems. The rise in fast fashion consumption which emphasizes "buy-use-throw away" leads to the release of microplastics into the environment [9].

Furthermore, these microplastics can absorb and accumulate toxic chemicals from the environment, which further amplifies the ecological risks. These microplastics are not only seen in the environment but also found in human traces like blood, lungs, breast milk, stools and even in air. These penetrate the various biological barriers causing issues in the lungs and gastrointestinal tract, etc. [10]. The release of textile microplastics has been raised from 2 billion metric tons to 8.3 billion metric tons from 1950 to 2017 and is anticipated to reach 34 billion metric tons by 2050 [11].

5 Waste Generation, Fast Fashion, and Consumer Waste in Textiles

The Industrial Revolution, growing consumerism, and developing nations significantly contributed to waste generation. One of the prominent sectors where waste generation is at its peak is the textile industry caused by the rise of fast fashion and increased consumerism. Out of 100 billion garments produced, 92 million tonnes end up in landfills and by the end of the decade, 134 million tonnes of clothing will be discarded. As a result of worn and thrown culture, clothes are worn only seven to ten times before being discarded. These tossed garments end up in a landfill but are not recycled [12].

Waste generation has become a global concern with an increase in demand for goods and services as a result of a consumer-driven society. This not only results in landfills of the garments but also the whole production process of the garment is involved in waste generation such as the textile lifestyle, raw material extraction to the production process, distribution, consumption, and disposal results in greater waste generation. In the initial phase, textile production involves various resources like water, energy, and raw materials. This process generates enormous amounts of waste including the release of hazardous chemical byproducts. Consumer behavior plays a vital role in driving the cycle of waste within the textile industry. The continuous change in styles and the low prices made the customers expect more for less, which in turn resulted in over-consumption and its subsequent disposal.

The environmental impact of this waste generation is multifaceted such as excessive resource consumption, and usage of hazardous chemicals contributing to various pollutions. The disposal of the textiles also leads to the release of greenhouse gases and toxins further aggravating climate change.

This waste generation can be addressed by various stakeholders. Manufacturers can adopt cleaner production methods, circular economy principles, implementing the concept of reduce, recycle and reuse, and usage of organic and sustainable materials can significantly reduce the industry's ecological footprints. The retail business models can focus on quality over quantity encouraging slow fashion concepts emphasizing classic designs and long-lasting products which can cut down the disposal. Consumers also play a role in driving change through mindful consumption, buying fewer and high-quality items that have higher life cycles can reduce waste disposal.

6 Waste Management—A Solution to Pollution

Waste management is the need of the hour to decrease the environmental pollution and impacts caused by the textile industry. Poor waste management has significant environmental health impacts, which lead to the need for effective waste management practices. This involves the methods and actions required to manage waste from beginning to end, including waste collection, transportation, treatment and disposal is crucial in the textile industry to reduce the impact and develop sustainability [13].

Manufacturers and brands play a major role in waste reduction by implementing sustainable practices. Implementation of lean production techniques, where the production is in line with the demand significantly reduces overproduction. Adopting digitalization can streamline the design process and minimize material waste. Brands working closely with the suppliers reduce energy consumption and can implement closed-loop recycling systems where the post-industrial waste is put back into the production process thus reducing the need for raw sources.

7 Need for Eco-textile

With the increased awareness of sustainability and environmental issues and ethics. Consumers are seeking products that align with sustainably produced and processed. Eco-textiles help in offering consumers socially and environmentally responsible products. Eco-textiles are the ones that are produced in an environmentally friendly way with less or no usage of chemicals and pesticides. These products help in reducing pollution. Since they are produced in sustainable methods, they help in conserving resources such as water and energy usage. This helps in reducing the greenhouse gas emissions and lower the carbon footprints. Since these textiles are free of harmful chemicals, they reduce their impact on humans and the environment. The reduction in environmental impact can be achieved with eco textiles as the resources used are natural and degradable. It helps in improving aquatic life as there will be a reduction in the long deposition of microplastics. As eco-textiles embrace the circular economy, it helps in the reduction of waste disposal thus reducing the textile landfills.

8 Government Policies and Regulation for Textile Pollution Control

The textile industry is the most polluting in the world, and the government has implemented several policies to curb the impact on the environment.

Various environmental standards have been established to control the pollution created by the textile industry. A few of them are the Global Organic Textile Standard (GOTS), EKO sustainable textile standard, Organic Fiber Standards (USA), Organic Textile Standard (Italy), and Standards for Processing of Organic Textile Products (Argentina). There are other rules and regulations to be followed while disposing of the wastes such as limitations to the pH, total suspended solids, biological oxygen demand of the effluent, azo-free dyes, reduced emissions of volatile organic compounds, use of organic fibres, minimizing carbon footprints, implementing processes to reduce the use of energy and water and avoid the discharge and minimal usage of organic solvents. Following these and having stringent action by the government help in reducing the environmental impact and threats caused by the textile industry [14].

9 Conclusion

In conclusion, the textile industry stands as a pivotal player in the global economy by meeting the growing need for textiles. However, it comes with a cost of environmental pollution and impacts that regulate climate change which cannot be ignored. Water pollution, created by the usage of harmful chemicals, endangers human and aquatic life. This requires the implementation of responsible chemical management, effluent treatment regulations, etc. to control water pollution. Another pressing concern is the energy consumption and emission of greenhouse gases. Transitioning to cleaner energy sources and implementing sustainable procedures help in reducing the impacts. Textile microplastics are often overlooked, which is often threatening to aquatic and human life.

Waste generation which is triggered by fast fashion and consumer culture adds to the landfills and greenhouse gases. This can be reduced by implementing the circular economy, reducing overproduction and promoting sustainable products and processes. Effective waste management practices are the need of the hour in reducing the textile footprints. The use of eco-textiles helps in reducing the environmental impacts of the industry. They help in implementing the usage of sustainably produced products encouraging the circular economy. These can be achieved with the intervention of government policies and regulations that are pivotal in controlling the textile industry impacts.

Ultimately, the textile industry is in phase to change contributing to the global positive change. By prioritizing sustainable and environmentally responsible products, it not only reduces the impacts but also contributes to a healthier planet for future generations and creates a massive change in the global climatic conditions.

References

1. https://textilevaluechain.in/in-depth-analysis/impact-of-textile-production-on-climate-change/
2. Choudhary MP, Islam S (2017) Assessment of environmental impacts during operational phase of a textile industry. Int Res J Eng Technol. www.irjet.net
3. Leal Filho W, Perry P, Heim H, Dinis MAP, Moda H, Ebhuoma E, Paço A (2022) An overview of the contribution of the textiles sector to climate change. In: Frontiers in environmental science, vol 10. Frontiers Media S.A.
4. Chioma (2023) The environmental impact of textile waste: 5 ways it impacts our planet. Faircado. https://faircado.com/mag/the-environmental-impact-of-textile-waste-5-ways-it-impacts-our-planet/
5. EP (European Parliament) (2018) Microplastics: sources, effects and solutions. https://www.europarl.europa.eu/news/en/headlines/priorities/fighting-plastic-pollution/20181116STO19217/microplastics-sources-effects-and-solutions
6. Whalen V (2022) Fast fashion and climate change 101. Action for the climate emergency. https://acespace.org/2022/06/17/fast-fashion-101/
7. Edwards O (2022) The importance of making textile industry energy-efficient—PCIAW®. PCIAW®—The Professional Clothing Industry Association Worldwide. https://pciaw.org/the-importance-of-making-textile-industry-energy-efficient/
8. Prince A (2008) Energy conservation in textile industries & savings. Fibre2Fashion. https://www.fibre2fashion.com/industry-article/3377/energy-conservation-in-textile-industries-savings
9. ECOS (2023) Ecos factsheet—microplastic pollution: the new runway trend for season 2023–2050. https://ecostandard.org/publications/ecos-factsheet-microplastic-pollution-the-new-runway-trend-for-season-2023-2050/
10. Euronews (2023) Microplastic pollution keeps getting worse. Filtration could be a fix. https://www.euronews.com/2023/04/24/microplastics-pollution-keeps-getting-worse-filtration-could-be-a-fix#:~:text=In%202023%2C%20most%20people%20understand,major%20causes%20of%20microplastic%20pollution
11. Periyasamy AP, Tehrani-Bagha AR (2022) A review on microplastic emission from textile materials and its reduction techniques. Polym Degrad Stab 199:109901. https://doi.org/10.1016/j.polymdegradstab.2022.109901
12. Igini M (2023) 10 concerning fast fashion waste statistics. Earth.Org. https://earth.org/statistics-about-fast-fashion-waste/

13. Kiron MI (2022) Waste management in textile and fashion industry. textilelearner. https://textilelearner.net/waste-management-in-textile-and-fashion-industry/
14. Fibre2Fashion (2014) Environmental standards for reducing pollution from textile and leather industry. https://www.fibre2fashion.com/industry-article/7287/environmental-standards-for-reducing-pollution-from-textile-and-leather-industry

Impact of the Textile Industry on Global Climate Change

Lata Samant, M. Pavan, Alka Goel, and Manpreet Kaur

Abstract The textile sector, which accounts for an important percentage of world trade, is being criticized for its numerous environmental issues. Numerous gases are being released by it, warming the Earth, making breathing more difficult, and damaging our water supplies. This occurs because the production of clothing consumes copious amounts of energy, water, and chemicals, and little fabrics wind up in the environment. This chapter examines in detail each stage of the textile production process and its impact on global climate. Utilizing meticulous techniques such as life cycle analysis is crucial to fully comprehend the implications. The study additionally examines the various types of pollution the textile industry produces and how they affect the environment and public health. The intricate nature of textile production is demonstrated. It also highlights how trash is produced, including chemicals and excess fabric, and how these actions may affect the ecosystem and human health in the long run. A thorough analysis of methods for sanitizing the water used to produce textiles is provided, aiding in the preservation of our aquatic life and flora. Governments, corporations, and common citizens must all cooperate to reduce the harm caused by the textile industry. In-depth knowledge of the ways the textile business impacts both our environment and humans is provided in this chapter. It is imperative that decision-makers, businesspeople, and scientists take note of this and act to improve the environment and public health.

Keywords Textile industry · Environmental impact · Climate change · Waste management · Life cycle assessment

L. Samant (✉) · A. Goel
Department of Clothing and Textiles, G.B. Pant University of Agriculture and Technology, Pantnagar, Uttarakhand, India
e-mail: samantlata100@gmail.com

M. Pavan
Department of Apparel and Textile Science, Punjab Agricultural University, Ludhiana, Punjab, India

M. Kaur
Department of Textile and Apparel Designing, University of Agricultural Science, Dharwad, Karnataka, India

Sadhna et al. (eds.), *Climate Action Through Eco-Friendly Textiles*, SDGs and Textiles,
https://doi.org/10.1007/978-981-99-9856-2_2

1 Introduction

As greenhouse gas emissions increase, global warming increases, and how the environment reacts to this warming is linked to it, which is one of the biggest issues facing our generation [1]. The amount of greenhouse gases (which trap heat) that are released into the atmosphere on a global scale and how responsive the Earth's climate is to those emissions determine how much the climate is changing. Rising sea levels, melting mountain glaciers, altered flower and plant blooming periods, severe drought, storms, heat waves, and an unending list of other big and little changes are all consequences of the trend toward higher global average temperatures. Since humans are the most sophisticated creatures capable of adapting to sudden changes, these changes are having much severe effect on the flora and fauna and the ecosystem that supports them. Being a link in the cycle, humans are also getting impacted both directly and indirectly at a slow and invisible pace. These changes are obvious and get more pronounced with each passing year.

A compilation report of the global rise in temperature from 1880 to 2020 has been published by NASA Earth Observatory, 2022, presented in Fig. 1. The observations from the Met Office Hadley Centre, the Berkeley Earth Research Group, the National Aeronautics and Space Administration, the NOAA National Centres for Environmental Information, and the Cowtan and Way study have been combined.

In order to bolster international efforts to combat climate change, one of the main objectives of the 2015 Paris Agreement and UN Climate Conference (COP21) is to keep temperature increases in this century below 2 °C over pre-industrial levels and below 1.5 °C. Furthermore, they intend to examine global commitments every 5 years and provide funding to developing countries so they can lessen climate change, increase resilience, and enhance their ability to adapt to its impacts [3, 4].

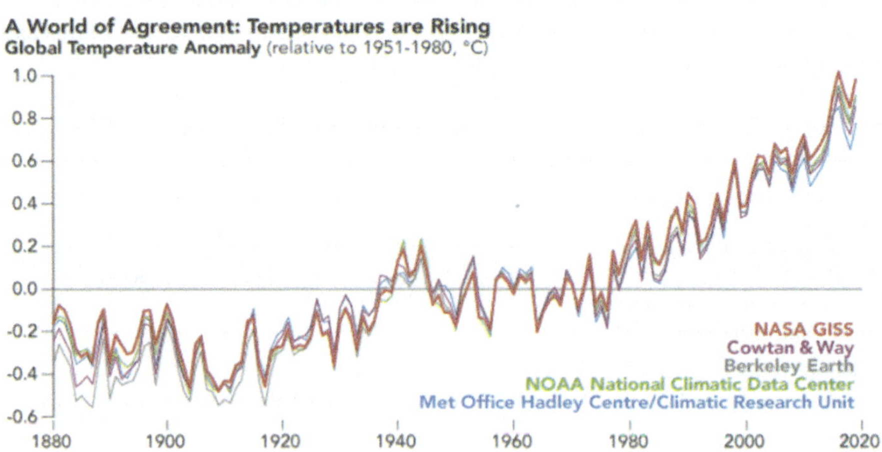

Fig. 1 Global temperature records from 1880 to 2020 [2]

2 Textile Industry CO_2 Emission

The textile and apparel industries are among the most polluting sectors of the economy because of the massive amounts of energy, water, and chemicals used, the creation of textile waste, and the release of microfibers into the environment [5]. The world's carbon emissions ranging from 8 to 10% come from the apparel and footwear sectors; these emissions outweigh those of the shipping and aviation industries put together [6]. This is a result of their intricate and vast supplier chains as well as their energy-intensive manufacturing processes. Furthermore, the dyeing and finishing of textiles may be responsible for up to 20% of industrial wastewater according to Morlet and team [7].

In an attempt to measure emissions across the board for the apparel industry, Quantis discovered that operations related to dyeing and finishing account for 36% of emissions, followed by yarn preparation (28%), fiber production (15%), and fabric preparation (12%) [8]. The other study, conducted by Sandin and colleagues [9], focused on emissions from six different types of clothing. It found that the majority of emissions (23.5%) came from wet treatment, which includes dyeing and finishing processes. Fiber production contributes to 16.3%, whereas waste from the cutting and sewing department and fabric production is 14.1%.

3 Carbon Footprint

The "carbon footprint" of an individual, group, activity, or product is the total amount of greenhouse gases released into the atmosphere, both directly and indirectly, especially carbon dioxide. It is estimated by calculating the carbon emission throughout the product's life or activity of an individual or organization. The globe experiences overload and temperature increase due to the CO_2 emitted from product activities that are trapped in the atmosphere [10].

4 Method of Estimating the Carbon Footprint

Input–output analysis and life cycle assessment are the most precise ways to determine one's carbon footprint. In addition to ISO 14040 and ISO 14043, ISO 14064 (parts 1, 2, and 3), ISO 14067 and PAS 2500 specify the carbon footprint of businesses and products [11]. Figure 2 shows the schematic model for determining a product's carbon footprint at various points throughout its life cycle.

Businesses and organization's greenhouse gas emissions fall into three categories: scope 1, scope 2, and scope 3 [13, 14]

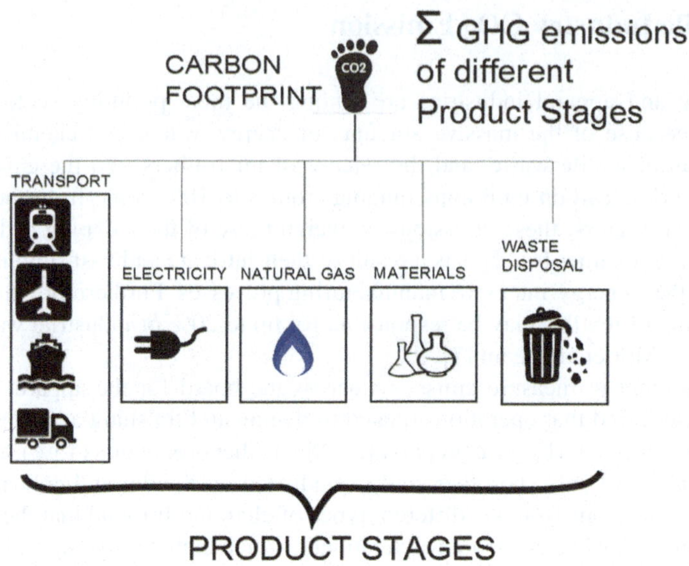

Fig. 2 Carbon emission estimation of the product life cycle [12]

(a) Scope 1—(Direct emissions) These are the emissions caused by the operations that the company owns or controls. It includes the use of machinery, vehicles, powering computers, etc.
(b) Scope 2—(Indirect emissions) Emissions generated by the company's acquisition of energy such as the organization's use of thermal energy, steam, or power.
(c) Scope 3—(Other indirect emissions) These are the greenhouse gas emissions apart from indirect energy. They are the GHG emissions from the company's operations that are consumed by others, such as goods that the company sells to suppliers and customers.

5 Models of Life Cycle Assessment (LCA)

The life cycle assessment (LCA) approach assesses the environmental impact of a product from the point of manufacture to the point of disposal. ISO 14040 and ISO 14044, standards developed by the International Organization for Standardization (ISO), are used for life cycle assessments. The "framework and principles" of the Standard are provided by ISO 14040, while the "requirements and guidelines" are outlined in ISO 14044.

As per the definition, the life cycle assessment is examining the life of material from its raw stage (cradle) till its disposal (grave) hence many times referred to as

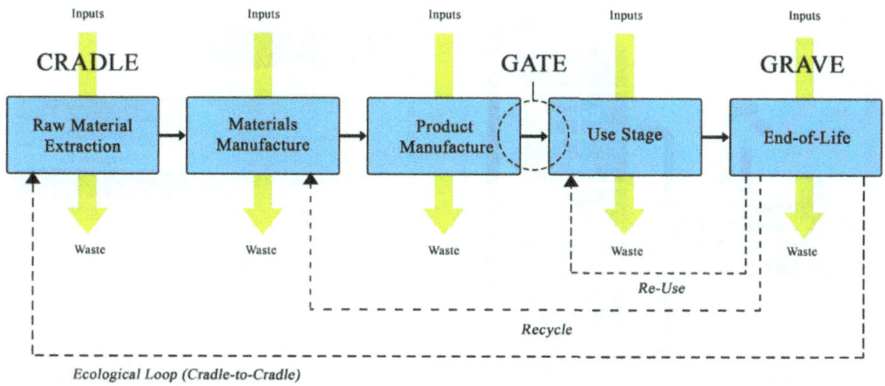

Fig. 3 Life cycle assessment of a product [15]

"cradle-to-grave analysis" the diagrammatic presentation of which is illustrated in Fig. 3 [15].

LCA models based on the different initial and ending stages the company is considering for the product.

- Cradle-to-grave: This method evaluates the effects of a product's life cycle at every turn, starting with extraction and continuing through each stage of manufacturing until final disposal.
- Cradle-to-gate: This is the carbon footprint of a product from the point of manufacturing to the point of entry into the factory gate (before transportation).
- Cradle-to-cradle or closed-loop production: It is a sustainable strategy where the product goes back to its natural state. It is a circular system in which the products get degraded back into the earth after use.
- Gate-to-gate: As the product moves from the production entryway to the departure gate, it evaluates its impact. This system does not cover the product's disposal phase.
- Well-to-wheel: It assesses an energy source's emissions and efficiency throughout its whole life cycle.

6 Impact of the Textile Industry on the Ecosystem

The most energy-intensive sector of the economy is textiles. Pretreatment, manufacturing, production, finishing, dyeing, printing, packaging, and distribution are just a few of the processing steps that go into making a garment that we wear. Gallons of water and energy are needed for each of these processes. The entire process's machinery is powered by fuel or electricity, both of which influence the environment. Additionally, a variety of chemicals are used during the preparation stage to transform the raw materials into a consumer-friendly final product of high quality. The energy,

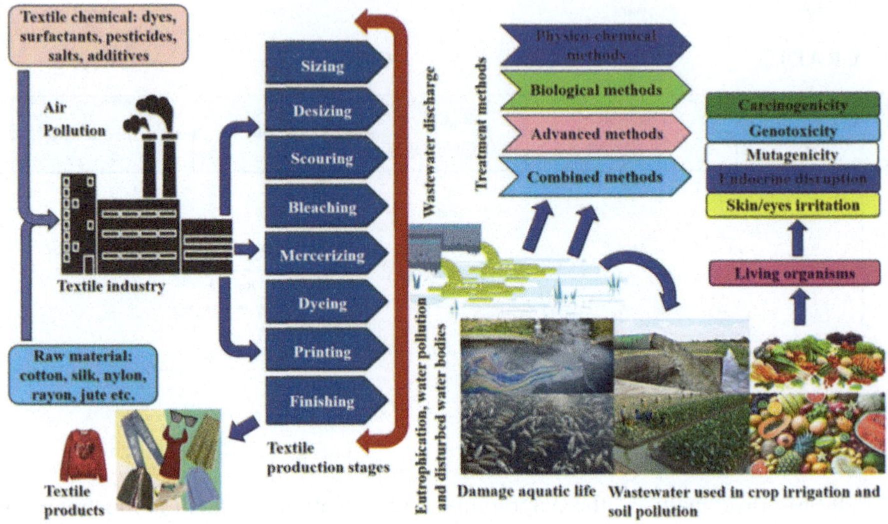

Fig. 4 Effects of the textile industry on the environment [16]

water, chemicals, and solid waste produced by the process unleash dangerous gases, water, and solid waste when discharged into the atmosphere, which can have many disastrous effects. Figure 4 provides a summary of the various phases of wastewater generation, its toxicity to human health and the environment, and available treatment options. Textile contributes to all the major environmental issues like air pollution (release of VOC, SSP), solid waste (fiber, scraps, leftover chemicals and resins, water pollution (chemicals, solvents, dyes), and noise pollution (heavy machinery). The textile industries and their raw material manufacturing units' pollution of the air, water, and land has become a major environmental issue. The effects of industry on people and the environment are listed below.

6.1 Water Pollution

Due to higher water consumption at various wet processing activities, the textile sector is an important producer of wastewater. Water is used extensively in almost all of the major steps involved in the manufacture of textiles. The processes of desizing, scouring, bleaching, mercerizing, dying, printing, and finishing are all part of wet fabric processing. Remains of the textile industry and chemicals are present in wastewater effluent. It contains a variety of ingredients, including colors, hydrogen peroxide, starches, surfactants, dispersions agents, acids, and metal-based soaps. It includes substances like hydrogen peroxide, starch, surfactants, dispersants, acids, alkalis, pigments, and metal-based soaps [17].

Aquatic life is endangered by wastewater dumped into bodies of water without proper treatment, and it is dangerous for humans and the environment that depends on it. It is the origin of many water-borne illnesses, including cholera, typhoid, hepatitis, and skin conditions. Though every stage of textile manufacturing releases liters of chemically contaminated water, the dyeing industry is the one that contributes the most to the generation of wastewater. Figure 5 shows the characteristics of wastewater released from various textile processes.

The different solvents, metal salts, sizing matter, dues, chemical, etc. effluents released from the textile operation pose a serious threat to the environment because they significantly reduce the oxygen concentration and obstruct light from penetrating the water body, which disturbs the water ecosystem. When waste is added to water, it raises the levels of total dissolved solids (TDS), biological oxygen demand (BOD), and total suspended solids (TSS). By absorbing light, these particles lower

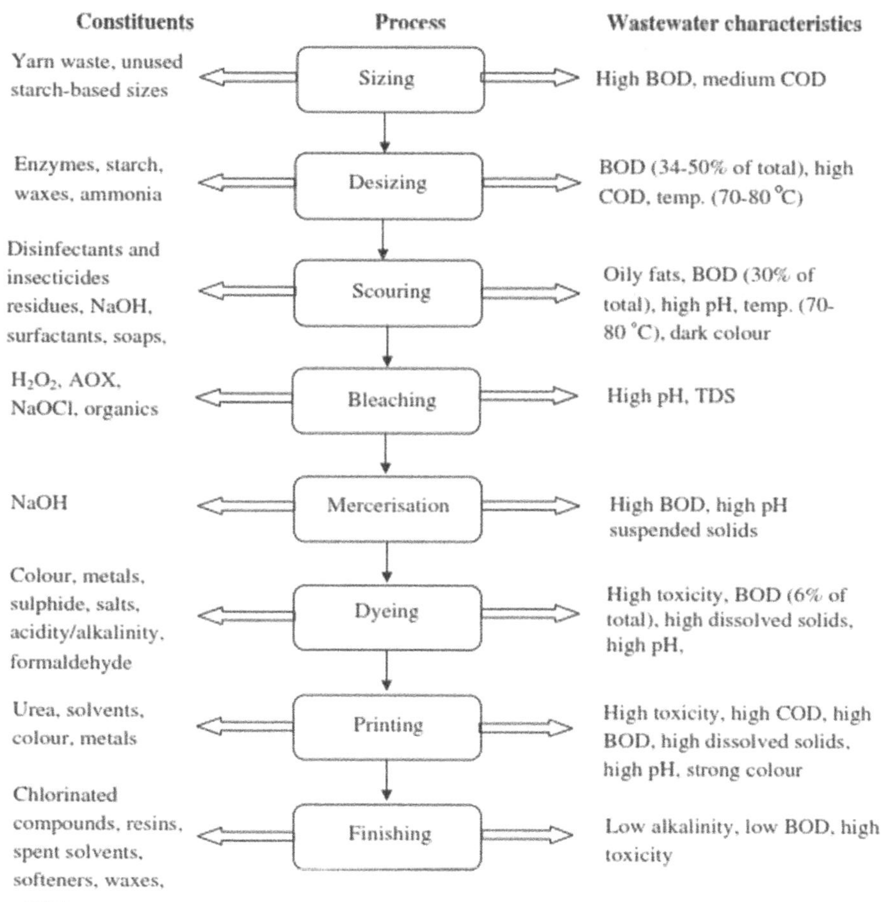

Fig. 5 Characteristics of wastewater released from various textile processes [20]

the dissolved oxygen content of the water and raise its temperature. This led to death of the many water creatures [18, 19].

6.2 Air Pollution

Another significant source of worry is the air pollution brought on by the textile sector. Due to the dangerous primary and secondary air pollutants that are generated, it pollutes both indoor and outdoor air. Suspended particulate matter (SPM), volatile organic compounds (VOCs), CO_2, aniline vapor, sulfur dioxide gas, oxide of nitrogen gas, etc. are only a few of the pollutants produced. The release of toxic chemicals into the atmosphere has had a detrimental effect on the world because air is a permeable substance that permeates the entire planet. The textile industry's emissions of pollutants must be drastically reduced. It has put the lives of humans and many other creatures on Earth at risk. The toxins that these industries release directly contribute to global warming. Individuals who are exposed to toxins over an extended period may potentially develop dangerous illnesses and health issues. The main effects of the pollutants released by industry are acid rain, ozone layer depletion, smog, mist, and others [21, 22]. Some of the major air pollutants released by the textile industry and their sources are given in Table 1.

Table 1 The major air pollutants released by the textile industry [23]

Process	Sources	Pollutants
Energy production	Emissions from boiler	Particulates, NO_x, SO_2
Coating, drying, curing	Emissions from high-temperature ovens	Volatile organic components
Cotton handling activities	Emissions from preparation, carding, combing, fabric manufacturing	Particulates
Sizing	Emissions from using sizing compound	Carbon monoxide, sulphur oxide
Bleaching	Emissions from using chlorine compound	Chlorine, chlorine dioxide
Dyeing	Disperse dyeing using carriers, sulphur dyeing aniline dyeing	H_2S, Aniline vapours
Printing	Emissions	Hydrocarbons, ammonia
Finishing	Resin finishing, heat setting of synthetic fabrics	Formaldehyde, carriers, polymers-lubricating oils
Chemical storage	Emissions from storage tanks for commodities and chemicals	VOCs
Wastewater treatment	Emissions from treatment tanks and vessels	VOCs, toxic emissions

6.3 Solid Waste

The need for fundamental amenities rises together with the pollution's exponential growth. Humans need clothing and textiles for survival, which increases trash production. Pre- and post-consumer waste are two categories of textile waste. Waste generated during the manufacturing process is included in pre-consumer waste [24]. Typically, the phases of a textile material are as follows: harvesting, spinning, fabric manufacture, cutting, sewing, yarning, dyeing, and packing. Solid trash produced includes packaging waste, fabric scraps, machine replacement parts, spools, microfibers, fabric scraps, chemical container waste, and more. Table 2 depicts the solid waste generated from the textile industry. People and the ecosystem will suffer long-term consequences if these wastes are not properly managed. The majority of solid waste is produced during the preparation of fibers, spinning of yarn, trimming and sizing, weaving, knitting, tufting, and finishing processes. Used clothing, towels, bedsheets, carpets, rugs, upholstery, and other textile products contribute to post-consumer trash.

Fast fashion (fad), buying power, and the influence of media have significantly increased the percentage of waste reaching landfills. It is an extremely hazardous contaminant that, if left unchecked, poses a serious risk to the continuation of life as we know it on Earth. In the long run, the chemicals from these garbage dumps

Table 2 Solid waste generated from the textile industry [21]

Process	Solid waste
Fiber preparation	Waste from packaging and fiber
Yarn spinning	Wastes from packing; sizing yarn; fiber; cleaning and processing waste
Slashing/ sizing	Fiber lint, waste from yarn, packaging waste, waste from cleaning, and unused starch-based sized
Weaving	Used oil, off-spec fabric, packaging waste, and yarn and fabric scraps
Knitting	Storing leftover yarn and fabric; off-spec cloth
Tufting	Storing leftover yarn, fabric, and off-spec fabric
Desizing	Packaging waste, lint from fibers, waste yarn, and supplies for upkeep and cleaning
Scouring	Minimal or none
Bleaching	Little or none; even if little, the impact could be considerable
Singeing	Minimal or none
Mercerizing	Minimal or none
Heat-setting	Minimal or none
Dyeing	Minimal or none
Printing	Minimal or none
Finishing	Trimmings and leftover fabric; packaging waste

seep into the ground and harm soil, crops, and aquatic life. These pollutants end up in landfills where they deplete land fertility and utilization [25, 26].

7 Effect on Human Health

Humans are directly or indirectly affected by the pollution caused by industries. The workers involved in different sectors of textile processing are the most vulnerable to the impact of chemicals, pollutants, and occupational hazards. They are at a high risk for the hazard. Whereas consumers are affected by the pollutant from the use of the product. A list of the risks associated with the textile business and how they affect human health is provided in Table 3.

Table 3 Risks associated with the textile industry and how they affect people's health [27]

Types of hazard	Causes	Effect on human health
Chemical hazard	Chemical agents releasing formaldehyde, and flame retardants are used in textile operations	Cancer, skin irritation, respiratory problem
Ergonomic hazard	Unsuitable workplace, improper ventilation, and lighting, awkward/ repetitive movements etc.	Musculoskeletal conditions include osteoarthritis of the knees, carpal tunnel syndrome, lower back discomfort, neck pain, and shoulder pain
Exposure to microfiber and dust	Excessive exposure, tightening of the chest, coughing, wheezing shortness of breath, etc.	Byssinosis, respiratory disorders, chest tightening, coughing, shortness of breath, etc
Physical hazards	Extreme temperatures, humidity, noise, hazardous radiation, fire, etc.	Body stressing, fatigue, irritation, headache, fluctuated heartbeat, hair impairment, etc.
Psychological hazards	Over time, the uninterested attitude of employees, low/ high workload demand, no social support, relations harassment and discrimination, etc.	Negative impact on mental health and well-being
Mechanical hazard	Equipment failure, low maintenance of machinery, etc.	Amputation of body parts, death, shock, paralysis, etc.

8 Textile Wastewater Treatment

The textile industry is sometimes classified into two groups: the wet and dry fabric industries, depending on the manufacturing process. Dry waste is created when the fabric is made into garments, cut, trimmed scraps, etc. On the other hand, liquid wastes are a byproduct of the cleaning or finishing operations that are used to treat the fabric. However, it is thought that wet processing produces more waste. The wastewater from the processing units contains chemicals that need to be treated before being released. Water treatment is more crucial since pollutants that can't be physically separated diffuse or disperse into the water. Wastewater contains a variety of pollutants, including ions, metals, colors, compounds that require oxygen, inorganic and synthetic organic molecules, and many others.

Numerous physiochemical mechanisms are involved in the release of hazardous waste from various textile enterprises. It is challenging to eliminate all of these contaminants with a single method. To purify the water, these treatments call for a set of steps to be followed. The three basic stages of treatment for textile waste are primary, secondary, and tertiary, as shown in Fig. 6.

8.1 *Primary Treatment*

It attempts to eliminate floating debris as well as settleable organic and inorganic substances by sedimentation. Using fine and bar screens, the coarse suspended particulates are taken out of the effluent. To get rid of the undesired effluents, many screenings are conducted. The remaining suspended particles are then removed by allowing the screened effluents to settle. Mechanical scraping methods are used to remove the

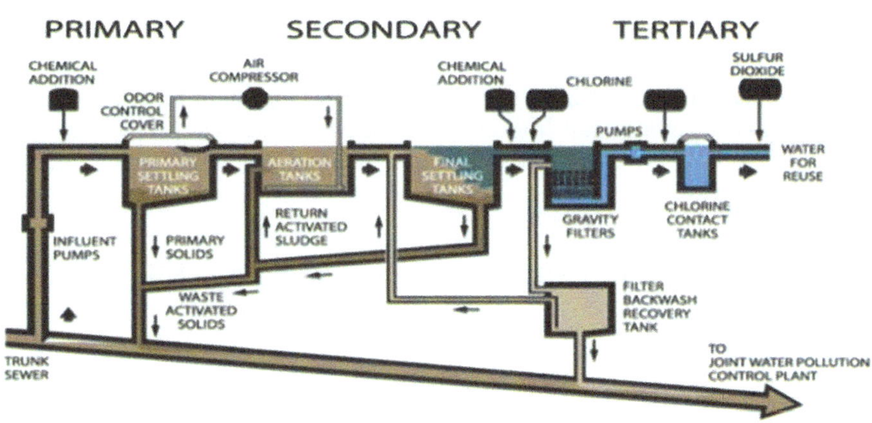

Fig. 6 Schematic representation of a typical Wastewater Treatment Plant [28]

floating debris. Neutralization is done to change the effluent's acidic pH. For the treatment process, a pH range of 5 to 9 is optimum [29–31].

8.2 Secondary Treatment

In secondary treatment, the biological mechanism plays a crucial role; it is primarily linked to lowering the effluent's color, BOD, COD, oil, grease, and phenol concentrations. Microorganisms can assist in physiologically accomplishing this under aerobic or anaerobic settings. Aerobic bacteria get their nourishment and energy from organic materials. They decompose nitrogenous organic compounds into ammonia and oxidize dissolved organic matter to produce CO_2 and water. Among the aerobic systems utilized in secondary treatment are activated sludge systems, trickling filters, and aerated lagoons. Stabilizing the generated sludge is the primary goal of anaerobic treatment. When oxidizing dissolved organic materials, aerobic bacteria frequently use it as a source of nourishment and energy. Ammonia, methane, CO_2, and water can be produced from nitrogenous and dissolved organic materials.

8.3 Tertiary Treatments

The last step in purifying wastewater is called tertiary wastewater treatment. It removes suspended particles, heavy metals, nutrients, and pathogenic organisms using a variety of physical and chemical techniques. Hazardous nutrients, suspended particles, and bacteria are removed frequently by filtering and then disinfecting. Various tertiary treatment strategies can be used, depending on the objective. Activated carbon adsorption, membrane separation processes (microfiltration, ultrafiltration, and reverse osmosis), ion exchange, disinfection (chlorination), and advanced oxidative techniques are a few of the methods that are most frequently employed. After treatment, the water can be recycled, safely used again, or released back into the environment, adding another level of defense against harm to people or wildlife [32].

9 Various Stages of Textile Production and Their Impacts on Society

In Table 4, a compiled summary of the textiles process is mentioned with the environmental and human hazards caused by it. The alternative chemicals used by industries are also mentioned.

Table 4 Various stages of textile production and its impacts on humans and the environment [33–36]

Process	Conventional chemical	Alternates	Human hazard	Environmental hazard
Singeing	A small amount of exhaust gas	–	Negligible impact	Textile effluents have high BOD, COD, total dissolved solids, total suspended solids, low dissolved oxygen values, and strong dye color that reduces the light penetration to the aquatic plants and animals, which affects their photosynthesis and overall ecological growth
Desizing	Enzyme, H_2SO_4, detergent, and alkali for PVA and CMC	Amylase enzyme	Bloating, diarrhea, and irritation to the skin	
Scouring	NaOH, Na_2CO_3, surfactant chlorinated solvent	Pectinase		
Bleaching	Hypochlorite, and acetic acid	Hydrogen peroxide	Chlorine gas release irritates the respiratory tract and eyes	
Mercerization	NaOH surfactant acid, liquid ammonia	Liquid ammonia	Irritation to skin	
Dyeing	Amino acid liberating group	Hydrogen peroxide, sodium perborate, pre-reduced dyes, and natural dyes	Dermatitis, nausea, skin and lung irritation	
Finishing	CH_2O, phosphorous, softeners fluorinated chemicals, catalysts, formaldehyde ammonia, paraffin, mineral salts, fluorocarbon resin	Plasma treatment, a combination of inorganic salts and phosphonates	Intense irritation to the eye and nose and headache, vomiting coughing, respiratory arrest, inhalation of harmful solvent may cause memory loss, confusion, affect the liver, asthma	

10 Conclusion and Future Outlook

One of the biggest sources of pollution and greenhouse gas emissions is the textile industry, which contributes significantly to environmental degradation. NASA's temperature records confirm how urgent it is to reverse this trend of increasing global warming. Initiatives like the Paris Agreement highlight the industry's critical role in reducing climate change and call for a shift toward sustainable practices. The textile industry is a significant carbon emitter, especially when it comes to dyeing and finishing. Hence, a critical evaluation of the carbon footprint of textile industries is essential, which is a crucial measure in determining the influence on the environment. Models of life cycle assessment, such as cradle-to-grave and well-to-wheel assessments, provide information on the environmental footprint of the textile sector.

Chemical-laden wastewater from various processing phases is a serious threat to aquatic environments and public health. At the same time, the burden of environmental degradation is increased by air pollution caused by volatile organic compounds and suspended particulate matter. Solid waste, both pre- and post-consumer, is becoming a more pressing problem, made worse by the spread of fast fashion trends. To mitigate long-term effects on land fertility and use, appropriate waste management measures are essential. The textile industry also poses a serious risk to human health because its employees are frequently exposed to chemicals and other contaminants.

A diversified approach is necessary to reduce the significant carbon emissions and environmental effects of the textile sector. First and foremost, it is critical to implement greener, more energy-efficient technology and procedures. Using cutting-edge dyeing methods, such as digital and waterless printing, can drastically reduce the amount of water and energy used. Furthermore, converting manufacturing processes to employ renewable energy sources can significantly lower greenhouse gas emissions. Furthermore, reducing trash output requires promoting circular economy concepts like recycling and upcycling. Using sustainable materials and implementing innovative eco-friendly production methods/practices can significantly decrease the industry's environmental impact. For the industry to be compliant and accountable, strict adherence to established environmental standards and laws is necessary, as are strong monitoring and enforcement systems. To drive systemic change toward a more sustainable textile industry, stakeholders—including governments, industry participants, and consumers—must collaborate. Last but not least, ethical and environmentally conscious brands should be supported. Proper care and maintenance can also extend the life of textile products, which can help to significantly reduce their environmental impact. Consumer awareness and responsible consumption practices also play a critical role in this.

References

1. Masson-Delmotte V et al (2021) Climate change 2021: the physical science basis. In: Contribution of working group I to the sixth assessment report of the intergovernmental panel on climate change, vol 2
2. NASA's Earth Observatory. https://earthobservatory.nasa.gov/world-of-change/global-temperatures. Accessed 1 Aug 2021
3. Sharma R, Sinha A, Kautish P (2020) Examining the impacts of economic and demographic aspects on the ecological footprint in South and Southeast Asian countries. Environ Sci Pollut Res 27:36970–36982
4. Abbass K, Qasim MZ, Song H, Murshed M, Mahmood H, Younis I (2022) A review of the global climate change impacts, adaptation, and sustainable mitigation measures. Environ Sci Pollut Res 29(28):42539–42559
5. Niinimäki K, Peters G, Dahlbo H, Perry P, Rissanen T, Gwilt A (2020) The environmental price of fast fashion. Nat Rev Earth Environ 1(4):189–200
6. Parliament European (2021) The impact of textile production and waste on the environment (infographic). European Parliament, Brussels
7. Morlet A, Opsomer R, Herrmann S, Balmond L, Gillet C, Fuchs L (2017) A new textiles economy: redesigning fashion's future. Ellen MacArthur Foundation
8. Quantis (2018) Measuring fashion: environmental impact of the global apparel and footwear industries study—full report and methodological considerations. Quantis 1–65
9. Sandin G, Roos S, Spak B, Zamani B, Peter G (2019) Environmental assessment of Swedish clothing consumption—six garments, sustainable futures. Mistra Future Fashion 1–164
10. Center for Sustainable Systems, University of Michigan (2021) Carbon footprint factsheet. Pub. No. CSS09-05
11. Szatyłowicz E, Skoczko I, Puzowski P (2021) Method of estimating the carbon footprint of wastewater treatment plants. Environ Sci Proc 9(1):30
12. Francesco F, Pietro B (2016) Carbon footprint as a tool to limit greenhouse gas emissions. INTECH 286–304
13. ISO (2013) ISO/TS 14067: greenhouse gases—carbon footprint of products—requirements and guidelines for quantification and communication, pp 1–52
14. Samantha S, Dominish E, Martinez-Fernandez MC (2022) Taking climate action: measuring carbon emissions in the garment sector in Asia (No. 53). ILO Working Paper
15. Klöpffer W (2008) Life-cycle based sustainability assessment of products. Int J Life Cycle Assess 13(2):89–94
16. Kishor R, Purchase D, Saratale GD, Saratale RG, Ferreira LFR, Bilal M, Bharagava RN et al (2021). Ecotoxicological and health concerns of persistent coloring pollutants of textile industry wastewater and treatment approaches for environmental safety. J Environ Chem Eng 9(2):105012
17. Paul SA, Chavan SK, Khambe SD (2012) Studies on characterization of textile industrial waste water in Solapur city. Int J Chem Sci 10(2):635–642
18. Zaw AK, Myat AM, Thandar M, Htun YM, Aung TH, Tun KM, Han ZM (2020) Assessment of noise exposure and hearing loss among workers in textile mill (Thamine), Myanmar: a cross-sectional study. Saf Health Work 11(2):199–206
19. Gleeson D, Labonté R (2020) Trade, labour markets and the environment. In: Trade agreements and public health. Palgrave Pivot, pp 93-112.
20. Cumnan S, Yimrattanabovorn J (2012) The use of constructed wetland for azo dye textile wastewater. Int J Civ Eng Build Mat 2(4):150–158
21. Karthik T, Gopalakrishnan D (2013) Impact of textiles on environmental issues. Part-II, Asian Dyer, pp 45–51
22. Rogers K (2018) What kinds of pollution do textile factories give off? Small Business, pp 1–17
23. Jutidamrongphan W, Rahman MA, Hossain T, Khatun SA, de Lamas W (2021) Eco-fashion designing to ensure corporate social responsibility within the supply chain in fashion industry. Autex Res J 21(4):467–481. https://doi.org/10.2478/aut-2020-0064

24. Mazharul IK (2022) Waste management in textile and fashion industry, Textile Learners. https://textilelearner.net/waste-management-in-textile-and-fashion-industry/
25. Sobhani A, Wahab MIM, Jaber MY (2019) The effect of working environment aspects on a vendor–buyer inventory model. Int J Prod Econ 208:171–183
26. Sumardiyono S, Wijayanti R, Hartono H, Probandari A (2019) The correlation between hearing loss and the quality of life of workers exposed to the noise in the textile industry
27. Sharma S, Babel R (2022) Safety and health issues in the textile industry fiber to fashion, health risks and protective measures for textile workers. Int J Multidisciplinary Educ Res 11(5):7–12
28. Sharma N, Bhatnagar P, Chatterjee S, Goswami P (2017) Waste water microbes and environmental "clean up": roadmap to environmental sustainability. Int J Adv Res Sci Eng Technol 4(2):3341–3350
29. Das S (2000) Textile effluent treatment—a solution to the environmental pollution. Fibre 2 Fashion
30. Babu RB, Parande AK, Raghu S, Kumar PT (2007) Textile technology-cotton textile processing: waste generation and effluent treatment. J Cotton Sci 11:141–153
31. Eswaramoorthi S, Dhanapal K, Chauhan D (2008) Advanced in textile waste water treatment: the case for UV-ozonation and membrane bioreactor for common effluent treatment plants in Tirupur, Tamil Nadu, India. Environment with People's Involvement & Co-ordination in India, Coimbatore, India
32. Singh S, Tiwari S (2019) Climate change, water and wastewater treatment: interrelationship and consequences. In: Water conservation, recycling and reuse: issues and challenges, pp 203–214
33. Pranaw MK, Tripathi MP, Sahu MR, Saxena MP (2014) Pollution of surface water in Tirupur and its adjacent areas by textile effluents. Centre for Science and Environment, New Delhi
34. Mathur N, Bhatnagar P, Sharma P (2012) Review of the mutagenicity of textile dye products. Univers J Environ Res Technol 2(2)
35. Pang YL, Abdullah AZ (2013) Current status of textile industry wastewater management and research progress in Malaysia: a review. Clean-Soil Air Water 41(8):751–764
36. Jeihanipour A, Karimi K, Niklasson C, Taherzadeh MJ (2010) A novel process for ethanol or biogas production from cellulose in blended-fibers waste textiles. Waste Manag 30(12):2504–2509

Eco-textiles—An Overview

Iti Dubey and Yogita

Abstract Based on the requirements of eco-friendly products, people are also aware of environmentally responsible manner textiles. This chapter is the collection of all data related to Oekotech. In addition to gain more of an understanding of eco textiles, this study is divided into two parts i.e., a brief description of the textile sector and the impact of eco textiles in terms of the global economy. We conclude that there are a lot of future outlooks for the industries working on eco-textiles as government norms and regulations also promote sustainable practices. This sector is primed for expansion, propelled by rising consumer demand, and products made from eco textiles are less harmful and made from biodegradable material.

Keywords Sustainable manner · Biodegradable · Global economy · Circular economy

1 Introduction

The word eco or oekotech textiles refers to textiles made in an environmentally responsible and sustainable manner. Eco textiles are produced with less harmful to the environment raw materials including organic cotton, hemp, and bamboo [1]. They are created utilizing techniques that employ the least amount of waste and pollution-causing chemicals, energy, water, and other resources. Eco-textiles also include recycled or upcycled materials. Recycled textiles are created from post-consumer waste, including used apparel and textiles, which are then treated and turned into new textile goods. Upcycled textiles are made from leftover pieces of fabric or production waste from the textile industry and are then used to make new items. Eco-friendly or oekotech textiles are produced using a variety of methods that utilize less water and toxic chemicals overall [2].

I. Dubey (✉) · Yogita
Department of Textiles and Apparel Designing, College of Community and Applied Sciences, Maharana Pratap University of Agriculture and Technology, Udaipur, Rajasthan 313001, India
e-mail: iti.dubey15@gmail.com

Sadhna et al. (eds.), *Climate Action Through Eco-Friendly Textiles*, SDGs and Textiles,
https://doi.org/10.1007/978-981-99-9856-2_3

Processes for dyeing and finishing textiles have also been developed to use fewer harmful chemicals and less water. Textiles that are eco- or oekotech have advantages for human health as well because they don't contain dangerous chemicals that can be absorbed via the skin [1]. In the textile industry, apparel consumption on the global level is on a rising track and it seems to continue increasing in upcoming decades and beyond [3]. This speed-up rate of production and consumption is leading to worse environmental impacts. Also, consumer scrutiny and tightening policy constraints already operating by the fashion and textile industry cannot be continued longer with business as usual. For economic development, decoupled from resource consumption, circular economy is an emerging approach. That can help society and regenerate the environment.

Overall, as knowledge of environmental issues rises, the necessity for eco- or oekotech fabrics becomes more critical [4]. By using sustainable and environmentally friendly production methods and materials, the textile industry may decrease its bad impacts on the atmosphere and amend to originate another bearable prospect. This chapter provides an overview of the need for and importance of eco or oekotech textiles and their impact on the global economy.

2 Importance and the Role of the Global Economy in the Textile Industry

The continuing production and trading of the planet's live resources for use as fuel, food, fiber, and feedstocks constitute a significant portion of today's global economy. However, the scientific community has issued a warning that changes brought about by humans are fundamentally altering the way our living world functions. Climate variability and environmental changes will have an impact on future resource flows. A change in productive regions will alter the political environment of commerce. The textile fashion sector is in the spotlight currently as societies all over the world mobilize for cleaner and greener products, and environmental change has taken the front stage in global politics [5].

The situation is getting more critical. Nations around the world endorsed 17 Sustainable Development Goals in 2015. These international agreements call for coordinated action on issues like climate change, biodiversity loss, and environmental protection, as well as many other undesirable and unsustainable effects brought on by the production and consumption systems of today. However, by 2020, uneven development continued, and the world was not on course to achieve the SDGs by 2030. The way the world responds to these planetary shifts is heavily influenced by business [6]. When compared to the long-term dynamics of environmental changes on a global scale, the timelines that matter the most to the business sector are often extremely brief. However, as risks increase and demands on planetary systems increase, society's capacity to respond quickly to change and mitigate the worst hazards depends on the trimestral and once-a-year planning cycles of business. It is time for businesses to make decisions for sustainability and resilience that consider the large-scale and long-term system conditions of the Earth.

3 The Need for Sustainable Alternatives

The traditionally produced textile industry is significantly harmful to the environment. The making of textiles involves a lot of resources, which include electrical power, water, and chemicals, affecting the environment and speeding up the rate of global warming. Additionally, while producing synthetic fibers like polyester, nylon, and acrylic, petroleum is a limited resource that must be mined and releases greenhouse gases. Traditional textiles are produced using hazardous chemicals, including dyes, bleaches, and finishing agents. There is a developing market for eco-friendly or oekotech textiles to offset these detrimental effects of the conventional textile industry on the environment and public health. These materials have a smaller negative environmental impact because they are biodegradable, renewable, or recycled [5, 7].

Moreover, the production processes used for eco or oekotech textiles involve sustainable practices that minimize waste, energy consumption, and pollution. Additionally, some eco or oekotech textiles are made using closed-loop production systems that recycle water, chemicals, and waste, easing the total excess produced and minimizing ecological pollution. In summary, the need for eco-friendly or oekotech textiles arises from the undesirable impression of the traditional textile trade on the environment and human health. Eco or oekotech textiles provide a sustainable alternative that uses environmentally friendly materials and production practices that minimize the impact on the environment and promote a sustainable future [8].

4 Global and Globular Economy

The global limits outline (Fig. 1) demonstrates by what means changes in personal behavior have altered Globe's geographical activities missing starting the Holocene epoch's 10,000-year steady baseline. The Earth's temperature and living ecosystem have been comparatively stable over this brief period of geological time, which made it possible for the establishment and growth of modern society. The hazards of disruptive environmental change increase as we move farther away from this "safe operating space." As a result of the present production and consumption trends, certain of the framework's processes are already under a lot of stress. Emphasizing these procedures identifies knowledge-based objectives for universal accountability on source usage and environmental repercussions [9].

The orbitual budget offers achievable recommendations for growth in the economy that benefits people and the environment. A circular economy restores natural systems, maintains goods and resources, and eliminates pollution and waste. The

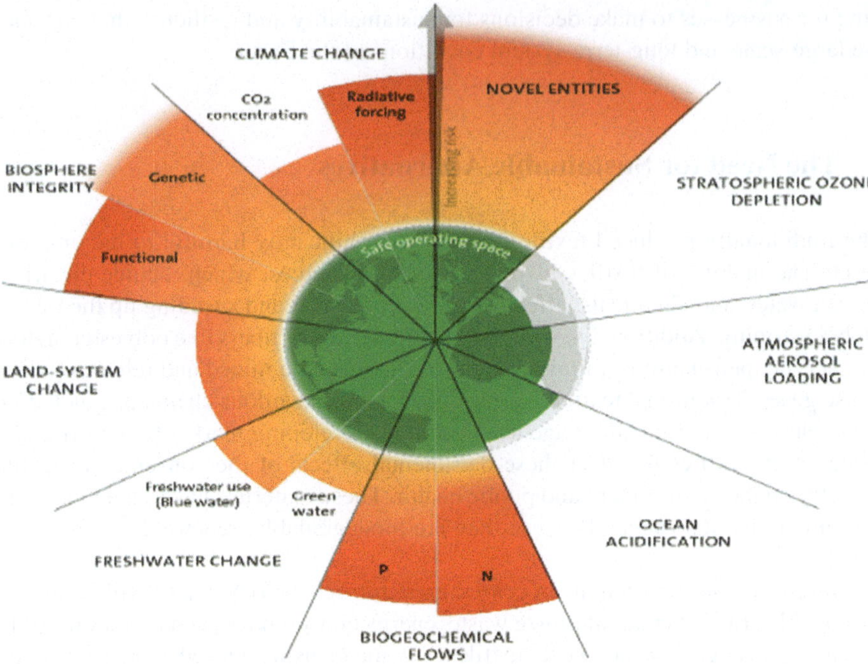

Fig. 1 The global framework shows the current earth system changes away from Holocene-like conditions [10]

circular economy's concepts and the global boundaries framework's aims may be used to provide a solid foundation for a regenerative and rejuvenating fashion and textiles industry.

5 The Principles of a Circular Economy

There are three guiding concepts for the circular economy model. Each addresses several of the resource and system issues that the textile system is currently facing or may encounter in the future [11].

5.1 Design Out Waste and Pollution

The damaging outcomes of commercial action that damage both personal healthiness and environmental methods are revealed and planned for by an orbitual reduction. This covers the discharge of glasshouse gases and theoretically unsafe mixtures, the damaging of the manner, ground, and rainwater in addition to underlying discarded like road traffic blocking.

5.2 Keep Commodities and Resources in Use

In a spherical budget, it is beneficial to participate in activities that save more cost in the establishment of force, effort, and resources. Maintaining commodities, mechanisms, and resources in the reduction requires designing for stability, recycling, remanufacturing, and recovering. Globular schemes effectively utilize organically built resources by hopeful multiple applications before returning nutrients to natural systems.

5.3 Regenerate Ordinary Arrangements

By recurring important nutrients to the earth to excite renewal or paying renewable energy in its place of vestige oils, a global budget minimizes the use of non-renewable possessions through preservatives or improving renewable ones [8].

6 The Fashion and Textile Industry is an Increasingly Complex System

The textile industry's sales, production, and solid resource consumption have all increased dramatically in recent years, and its effects on society and the environment have spread throughout the world (Fig. 2). This has led to complex relationships and domino effects between the industry, its many stakeholders, and the natural world. If the industry doesn't quickly change its practices, it will continue to contribute to social issues and environmental consequences from a systemic perspective.

Many sector projections predict that the textile and clothing sector will expand during the coming decades. They project existing movements into the imminent in the hope that a greater, affluent, and healthful worldwide customer will be interested in purchasing fashion clothing. However, as the sector grows internationally, it also takes more resources from the living world and requires more feedstock based on fossil fuels. The planet's ability to maintain supplies of raw materials and absorb polluting emissions may hinder industry expansion [8].

Fig. 2 The fashion and textiles industry in numbers [12]

7 Problem Dynamics Lead to Rising Global Pressures

The textile industry and governments, considering these ongoing environmental challenges, advocate switching from today's circular economy to a circular economy by reusing precious textiles and generating new vital goods. Making this transition requires better robust recognition of how modern technology uses waste style and fabric classifications to impact changes in planetary-ruler schemes. Table 1 lists the ways in which industry contributes to worldwide pressures. Industry's increasing contribution to planetary stress is driven by three important reasons.

- Production and consumption of textiles have increased faster than anticipated by the industry. In a linear economy, the strain on the planet's natural resources increases as an industry expands more quickly.
- Demand from consumers has increased. The economic pressure to produce and sell more items has led to a trend towards less durable and lower quality goods, worsening the pollution and waste issues.

Today's textile and fashion industries are bound by unsustainable patterns of behavior. Organizations' resource allocation and material flow management are impacted by a variety of beneficial, technical, prejudiced, and intellectual aspects of national and international economies [11].

In the fashion and textiles system, these elements have worked in concert to produce a self-reinforcing loop where production, consumption, and waste leakages have all increased at the same time.

Table 1 Industry's contribution to global pressures [12]

	Environment transformation: all artificial fiber manufacturing, liveliness consumption, and fabric transference involve the use of vestige oils. The sector produces 10% more carbon dioxide than the 1% contributed by fiber production alone
	Biodiversity loss: threats to life along the value chain include soil erosion, deforestation (and CO_2 emissions), monoculture farming, introduction of non-native species, midair, water, and soil effluence for all fibers
	Land-living revolution: small-scale farming is no longer possible in some environments due to soil erosion, deforestation, desertification, and soil salinization, which has led to dramatic increases in cotton production
	Freshwater use: all manufacturing involves the use of water. Water availability affects the yield of crops used for fiber. Intensive irrigation causes agricultural fields to become eroded and saline, increasing the need for fresh water
	N and P fluxes: typically, chemically intensive agriculture loses nutrients, causing eutrophication and increasing nitrous oxide emissions, which deplete the ozone layer and are greenhouse gases. Eutrophication also causes nutritional deficiencies
	Chemical pollution: toxic chemicals used in today's fibers and textiles, including dyes, treatments (such as water and stain repellents), pesticides and other agricultural chemicals, polluted runoff, and waste

In addition to pushing the system toward a state that is undesirable from a social, ecological, and economic perspective, these dynamics have also contributed to a much more serious technological and institutional lock-in scenario.

Conversely, ethical businesses must understand their responsibilities by considering not only the immediate effects of an invention's lifetime but also the eco-friendly influences symbolized in the global skill of raw materials and the ingesting habits of style consumers. Let's try to make it comprehensive [13].

8 Universal Aims to Set the Framework for Achievement

Global eco-friendly requirements that address global issues provide an essential, long-term basis for occupational involvement in the conversion to a global cheap. The goals of these multinational ecofriendly contracts are stated in the context of future states, extending their strategic timelines beyond the next 10 years. Countries have arranged that attaining these aims underwrites a supportable and strong future.

These objectives are guided and observed by expert skill and program societies. With these goals in mind, each of the six planetary priorities could have specific global environmental targets. To avoid short-term measures on one front harming long-term progress on other planetary challenges, it is necessary to constantly keep these long-term global goals in mind [13].

8.1 In Brief

- By 2050, the global goal for combating climate change is to achieve carbon neutrality. This is the outcome of the Paris Agreement's aspirational goal of limiting global warming to 1.5 °C above pre-industrial levels.
- To combat biodiversity loss, the world's population must conserve 30% of the planet's land area through conservation measures that protect human rights. In contrast to controlled degradation, the world is now on the path to regeneration. According to the 2050 vision of the Agreement on Organic Range, biodiversity must be successfully reinstated, appreciated, and cast off to ensure a healthful earth and deliver essential profits for all people.
- The worldwide aim of land use is to reinstate 20% of the Earth's property surface to an organically sound condition. It also helps prevent land degradation and desertification and realize the 2050 vision of the Convention on Biological Diversity. When done carefully, it can also help prevent disruption of the hydrological cycle in nearby areas and mitigate the effects of climate change.
- The overall aim of river use is to keep freshwater removals in all crises below 40% of renewable resources. It contributes to efforts to avoid extreme stages of water strain, decrease increasing hazards of water shortage, and mitigate the effects of interrupted hydrological sequences dispersion to other regions.
- Although there does not appear to be a single international agreement governing water, there is broad agreement among scientists and strategy makers that the rapid increase in river use realized since the 1950s must be ceased and, if possible, it must be upturned.
- Applying source containment, discharger pays, and protective values helps change manufacturing to fully spherical and regenerative organizations. Detailed chemicals and submissions are also subject to limits or restrictions below several international agreements.
- For a circular economy to support a sustainable and resilient society, global impartiality should have taken into account preventing pollution issues before they become hazardous damages rather than in the context of addressing issues later on after they have already caused significant harm.
- N&P Flux's worldwide aim is to increase long-term full-shackle nutrient consumption productivity by 50% for all bio-constructed source construction. Despite no single global consensus on managing nutrients in water, there is broad agreement about how to reverse increasingly dangerous tendencies of wasteful use, ecological leaks, and publicly high outwardness. It takes a very long time, and it's not easily understood how to refurbish biological stability next excessive

enrichment of N and P. In general terms, releasing N and P into the environment is an unnecessary waste of natural resources [14].

Some of the emissions, effluents, and solid wastes produced by various textile production processes are given in Table 2.

9 Innovations in Eco or Oekotech Textile Production

Innovations in eco or oekotech textile production are driving the industry toward more sustainable and responsible practices. Here are a few notable innovations.

9.1 Sustainable Fiber Alternatives

Development of plant-based fibers like organic cotton, hemp, and bamboo that require fewer resources and chemicals.

9.2 Recycled and Upcycled Materials

Transforming waste materials into new textiles, such as polyester fibers from plastic bottles, and upcycling discarded fabrics.

9.3 Low-Impact Dyes and Printing Methods

Eco-friendly dyes from natural sources and digital printing techniques that minimize water usage and chemical waste.

9.4 Waterless and Energy-Efficient Manufacturing

Processes that reduce water consumption and energy usage in textile production.

Table 2 Textile industry waste management [14]

Process	Emission	Wastewater	Solid waste
Fiber preparation	Minimal or none	Minimal or none	Waste from fiber and packing
Yarn spinning	VOCs	Minimal or none	Sized yarn, cleaning, and manufacturing waste, and fiber and packaging waste
Slashing/ sizing	VOCs	Metals; biochemical oxygen demand (BOD); chemical oxygen demand (COD)	Waste from cleaning, packaging, and yams, as well as wasted starch-based sizes
Weaving	Minimal or none	Minimal or none	Scraps of fabric and yarn, as well as packaging and spent oil
Knitting	Minimal or none	Minimal or none	Packaging trash, off-spec fabric, and yarn and fabric scraps
Tufting	Minimal or none	Minimal or none	Fiber lint, yam waste, cleaning waste, off-spec fabric, and packaging trash
De-sizing	Glycol ether-derived VOCs	Size-based BOD from lubricants, biocides, and antistatic agents	Materials for cleaning and maintenance, fiber lint from clothing, and yam waste
Scoring	VOCs produced by scoring solvents and glycol ethers	Knitting lubricant, pesticide remnants, detergents, oils, spin finishes, and wasted solvent	Minimal or none
Bleaching	Minimal or none	Stabilizers, high PH, and H_2O_2	
Singeing	Small quantities of burner exhaust gases	Minimal or none	Minimal or none
Mercerizing	Minimal or none	NaOH and high PH	Minimal or none
Heat setting	Manufacturing of synthetic fibers using the volatilization of spin finishers	Minimal or none	Minimal or none

(continued)

Table 2 (continued)

Process	Emission	Wastewater	Solid waste
Dying	VOCs	Metals, salt, surfactants, alkalinity, acidity, color, BOD, COD, used organic solvents, and processing aids	Minimal or none
Printing	Chemicals, acetic acid combustion gases from drying and curing oven emissions	Color, metals, heat, foam, Urea solvents, suspended solids, and BOD	Minimal or none
Finishing	Pollutants and volatile organic compounds formaldehyde vapors and combustion gases in chemicals bought	COD, suspended solids, used solvents, and hazardous substances	Packaging trash; fabric trims and remnants

9.5 Nanotechnology Applications

Enhancing textile properties with nanoparticles for durability, water repellence, and UV protection [9].

9.6 Smart Textiles and Wearable Technology

Integrating electronics and sensors into textiles for health monitoring and functional benefits [9].

9.7 Blockchain and Supply Chain Transparency

Using blockchain technology to track and ensure ethical sourcing and production practices.

9.8 Sustainable Finishing Processes

Enzyme-based treatments and natural resins replacing chemical finishes for improved environmental impact [2, 3].

These innovations reflect the industry's commitment to sustainability, social responsibility, and meeting the demand for eco-friendly textiles. They contribute to reduced environmental impact and improved production practices throughout the textile value chain.

10 Opportunities for Growth and Development in the Industry

The eco or oekotech textile industry presents several opportunities for growth and development. The following are a few of the major changes.

10.1 Increase in Consumer Demand

Because consumers are increasingly requiring sustainable and ecologically friendly products, including textiles, the textile industry can meet these expectations.

10.2 Innovation in Sustainable Materials

By developing new sustainable materials like bio-based fibers and recycled textiles, the industry may increase the range of products it offers and adapt to shifting consumer preferences.

10.3 Technological Advancements

Embracing technology, such as digital printing and automation, can optimize production processes, reduce waste, and enhance efficiency in eco or oekotech textile manufacturing.

10.4 Collaboration and Partnerships

Collaborating with industry stakeholders, researchers, and designers fosters innovation, knowledge-sharing, and the development of sustainable practices and products.

10.5 Certifications and Standards

Adoption of standardized certifications and labeling for eco or oekotech textiles enhance credibility, transparency, and consumer assurance.

10.6 Policy Support

Supportive government policies, such as incentives and procurement preferences for sustainable textiles, create a favorable business environment and drive industry growth.

10.7 Circular Economy Practices

Embracing circular economy principles, like recycling and upcycling, minimizes waste, reduces resource consumption, and opens new business opportunities.

10.8 Ethical Fashion and Slow Fashion Movement

Aligning with ethical fashion principles and the slow fashion movement, which focuses on the durability and timeless designs, allows the industry to tap into the growing consumer interest in socially responsible fashion [15].

By capitalizing on these opportunities, the eco or oekotech textile industry can experience growth, market differentiation, and increased profitability, while contributing to environmental sustainability and responsible consumption.

11 Conclusion

In conclusion, through textile processing and production both units reported that there has been too little Biochemical oxygen demand and chemical oxygen demand for metals and water waste. There is some solid waste too in processing units, i.e., fiber waste, packaging waste, cleaning, fiber tint, etc. These conditions are very serious and affect environmental industries that contribute to planetary pressures. Apart from this, most of the sectors estimate anticipation that the clothing and textile industries will expand during the coming decades. However, as the sector grows internationally, it also takes more resources from the living world and requires more feedstocks based on fossil fuels. This sector has faced lots of problems, but growth and new developments take this sector to new heights. Additionally, some eco or oeko textiles are made using closed-loop production systems that recycle water, chemicals, and waste, reducing the amount of waste produced and minimizing environmental pollution.

References

1. Aishwariya S (2019) Inclination towards Oeko-textiles (Jayakumar R (ed)). Res Trends Multi Res 1:160–176
2. Nayak R, Panwar T, Nguyen LVT (2020) Sustainability in fashion and textiles. Sustain Technol Fashion Text 3–30. https://doi.org/10.1016/b978-0-08-102867-4.00001-3
3. Aldalbahi A, El-Naggar M, El-Newehy M, Rahaman M, Hatshan M, Khattab T (2021) Effects of technical textiles and synthetic nanofibers on environmental pollution. Polymers 13(1):155. https://doi.org/10.3390/polym13010155
4. Cornell S, Häyhä T, Palm C (2021) A sustainable and resilient circular fashion and textiles industry. The sustainable textiles project, pp 7–10
5. Gulzar T, Farooq T, Kiran S, Ahmad I, Hameed A (2019) Green chemistry in the wet processing of textiles. The impact and prospects of green chemistry for textile technology, pp 1–20. https://doi.org/10.1016/b978-0-08-102491-1.00001-0
6. Environmental, Health, and Safety Guidelines for Textile Manufacturing. International Finance Centre, Central, Hong Kong (2007)
7. Lee KE (2017) Environmental sustainability in the textile industry. In: Muthus S (ed) Sustainability in the textile industry. Textile Science and Clothing
8. Rathinamoorthy R (2019) Circular fashion. In: Circular economy in textiles and apparel. Elsevier, pp 13–48
9. Kaounides L, Yu H, Harper T (2007) Nanotechnology innovation and applications in textiles industry: current markets and future growth trends. Mater Technol 22(4):209–237. https://doi.org/10.1179/175355507x250014
10. University of Copenhagen (2023) Six of nine planetary boundaries now exceeded. YubaNet. https://yubanet.com/scitech/six-of-nine-planetary-boundaries-now-exceeded/
11. Kooistra K (2006) The sustainability of cotton. Science Shop Wageningen University and Research Centre, p 223
12. Cornell S, Gomėz PV, Kallstenius I, Hileman J, Palm C, Häyhä T (2021) A sustainable and resilient circular fashion and textiles industry. Stockholm Resilience Centre. https://www.stockholmresilience.org/research/research-news/2021-04-12-six-targets-for-a-sustainable-textile-industry.html
13. Henry B (2019) Microfibres from apparel and home textiles. Sci Total Environ 652:483–494
14. Niinimäki K (2020) The environmental price of fast fashion. Nat Rev Earth Environ 1(2020):189–200
15. Scaturro S (2008) Eco-tech fashion: rationalizing technology in sustainable fashion. Fashion Theor 12(4):469–488. https://doi.org/10.2752/175174108x346940

Role of New-Generation Textile Fibres in Reducing the Environmental Impact of Textiles

Meenakshi Tamta and Arpana Kamboj

Abstract New-gen or new-generation textile fibres transformed the textile industry to a different level through the sustainable and eco-friendly characteristics of these fibres. These fibres can be obtained from different renewable resources, which offer several benefits such as dependency on non-renewable resources, reduced carbon footprint and improved waste management. These new-gen fibres have several applications in different areas such as technical textiles, geotextiles, protective clothing and filtration membranes. These new-gen fibres are facing many challenges such as production process, scalability, processing cost and awareness about these fibres among consumers. After having all these issues these fibres have a promising and great future. With the rising consumer awareness about sustainability, demand for these fibres is growing day by day. These new-gen fibres play a significant role in shaping sustainability and are responsible for a bright future for the textile industry. So, new-generation textile fibres provide a pathway to a greener future, contributing to environmental conservation, resource efficiency and social well-being. Embracing these fibres and promoting sustainable practices throughout the textile supply chain are crucial steps towards creating a more sustainable and circular economy.

Keywords Apparel · Greener future · Renewable resources · Sustainable

1 Natural Textile Fibres

The fibres from natural materials can be obtained from plants or animals and can be used for the manufacturing of textiles and other items. Mankind has been using natural fibres for several thousand years and they continue to be preferred due to their desirable characteristics and eco-friendliness. In accordance with their manufacturing, fibrous items raw materials can be divided into two basic types that are natural and synthetic. Natural fibres have been there since the beginning of time, while man-made fibres have only recently been invented [1]. Natural textile fibres,

M. Tamta (✉) · A. Kamboj
Swami Vivekanand Subharti University, Meerut, Uttar Pradesh 250002, India
e-mail: meenakshitamta04@gmail.com

Sadhna et al. (eds.), *Climate Action Through Eco-Friendly Textiles*, SDGs and Textiles,
https://doi.org/10.1007/978-981-99-9856-2_4

as the name signifies, are derived from sources that are naturally occurring sources. Cotton, jute, hemp, ramie, pineapple, and linen are examples of naturally occurring fibres derived from plants. The structural component of natural fibres is cellulose and it is an abundantly available polymer found in nature [2]. These natural cellulosic fibres have high strength, medium to light in weight, low priced, abundantly available in nature and renewable in nature [3]. The natural fibres have adverse effects of some of the factors such as climatic conditions, and variety of the fibre.

Many factors influence the qualities of natural fibres, including selection, environmental circumstances, harvesting, maturation, retting, decortications, fibre alterations, textiles and technological processes (spinning and carding) [4]. Attributes such as dimension, durability, pliability, flexibility, resistance to abrasion, water absorption and other surface qualities define a natural fibber's commercial utility. With all of these desirable features, natural fibres can be converted into woven fabric once the yarn is spun and into non-woven fabrics like paper and felt by applying pressure as well as heat [5].

Natural fibre has become a significant alternative to synthetic fibres due to its abundance, low cost, biodegradable and recyclable nature and great durability. The usage of natural fibres will solve several environmental issues. Natural fibres utilising conjugated polyester matrix outperform those with unreinforced polymers in terms of both composite strength and toughness. Furthermore, cellulose natural fibres are suitably strong, lightweight, inexpensive, plentiful, and sustainable [3]. These also have additional benefits including accessibility at inexpensive prices, non-toxicity, non-carcinogenicity, recycling capacity and many more. The durability and ability to absorb water of such fibres are primarily determined by their constituents, which include hemicelluloses, cellulose lignin, pectin and wax. Cellulose gives fibres durability, but hemicellulose enhances the absorption of water capacity [6]. Natural fibre material is used in a range of occupations, notably automobiles, articles of furnishings, wrapping, and buildings. That's primarily due to its benefits above fibres made from a synthetic polymer which include inexpensive prices, flexibility, lower damage to production machines, improved surface polish of composite components, excellent relative mechanical properties, and plentiful and renewable energy sources [7]. Fibres made from nature are utilised in a variety of uses, including building supplies, particle panels, insulating surfaces, nutritional food and livestock feed, pharmaceuticals, cosmetics, and other biological polymers and synthetic chemicals [8].

2 New-Generation Natural Fibres

New-generation natural fibres refer to innovative and advanced types of natural fibres that have been developed through various techniques and processes apart from conventional methods (Fig. 1) to enhance their properties or create entirely new fibre materials. These fibres often combine the benefits of natural materials with improved performance characteristics, making them suitable for many applications.

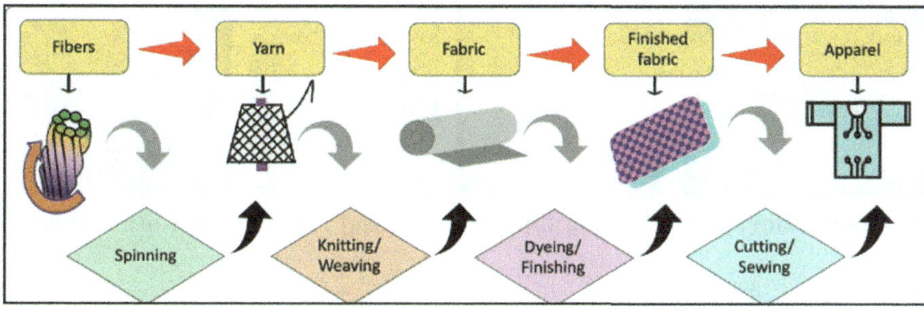

Fig. 1 Manufacturing process from fibre to fabric of new-gen fibres [11]

Natural fibres have shown tremendous promise in several applications that were previously dominated by synthetic fibres due to the critical features of biocompatibility, being biodegradable non-toxicity, and availability [9]. Next-gen natural fibres are an environmentally friendly substitute to petroleum-based plastic since they are derived from renewable resources that promote commercial composting at the end of their lives. The need for sustainability is at an all-time high in the fashion industry, for example, and many companies are attempting to produce more next-gen environmentally friendly fibres and materials. Natural fibres are made from renewable resources such as cellulose from plants and protein from animals. Cotton, jute, sisal, coir, flax, hemp, abaca, ramie and other natural cellulosic fibres are included, as are protein-based fibres such as wool and silk. Certain crops can produce multiple types of fibre. Jute, flax, hemp, and kenaf, for example, are found in both bast and core fibres, whereas agave, coconut, and oil palm include both fruit and stem fibres. Grain cereals also include stalk and shell fibres [10].

These new generations of natural fibres offer alternatives to synthetic materials and contribute to the development of sustainable and environmentally friendly products. They often provide unique characteristics and benefits that make them attractive for various industries, including fashion, textiles and home goods.

Here are a few examples of new-generation natural fibres

Bamboo—Bamboo fibre is derived from the bamboo plant and is considered a sustainable and eco-friendly alternative to traditional fibres. Bamboo is a Bambusoideae grass composed of cellulose fibre encased in a lignin structure. Bamboo has several advantages over other plant-based fibres, including low density, low cost, excellent structural rigidity, rapid expansion and the ability to fix atmospheric CO_2. Bamboo has traditionally been used in building and the development of items for daily living due to its high strength-to-weight ratio. It is suitable for clothing, bedding and towels because of its natural antibacterial and moisture-wicking properties [12].

Banana—Banana fibre (Fig. 2a) is derived directly from the banana plant's shoots. The banana is one of the oldest cultivated plants. It is a member of the Musaceae family [13]. The fibre of bananas is a lingo-cellulosic fibre derived from the pseudo stem of the banana plant. It has natural lustre and moisture absorption properties. It

Fig. 2 Some new-gen fibres [20–23]

is a tough and long-lasting fibre that can be found in clothing, cords, placing mats, paper cardboard, rope yarn, bags for tea, superior fabrics for textiles, cash notes paper, cables, cushion covers, table top cloth, drapery, thin-walled composite materials, and biological fertiliser [14].

Areca—Areca fibre, also known as areca palm fibre or betel nut fibre is derived from the outer husk of the areca palm tree (Areca catechu). The Areca husk is a strong fibrous layer that protects the endosperm from damage. Hemicellulose is the main constituent of these fibres. Areca fibres have a chemical structure consisting of lignin (13–24%), hemicellulose (35–64.8%), ash (4.4%), and water content (8–25%) [15]. This fibre is primarily used for various non-textile applications, such as packaging, handicrafts and biofuel. It can also be mixed with cotton or polyester to add strength, durability, and a unique texture to the fabric.

Oil palm—Oil palm fibre, commonly known as palm fibre or palm mesocarp fibre, is obtained from the oil palm fruit's mesocarp. It is the world's high-yielding commodity of edible oil. It is grown in 42 nations on 11 million hectares of land globally. The primary oil palm-producing nations are Southeast Asian countries such as Malaysia and Indonesia, West Africa, Latin American countries and India [16]. Oil palm is primarily used in non-textile applications such as mattress filling, erosion control and biofuel, there are also some potential uses of oil palm fibre in the textile industry such as yarn Production, upholstery and padding, geo-textiles, eco-friendly textiles and decorative applications.

Pineapple leaf fibre—Pineapple leaf fibre (Fig. 2b) is derived from the outermost layer of leaves of the Bromeliaceae—family shrub *Ananas comosus*. The pineapple plant's leaves produce robust, white thin smooth fibres. Because the plant that produces pineapples is a type of speciality harvest and from this crop limited amount of fibre is accessible [17]. Pineapple fibre is commonly used to produce a luxurious textile known as piña silk. It can also be mixed with other fibres, like silk or cotton, to create blended fabrics with enhanced properties. Home textile applications of pineapple fibre include curtains, upholstery, tablecloths, and decorative fabrics.

Sisal fibre—Sisal fibre (Fig. 2c) is an extremely tough fibre. It is derived from the *Agave sisalana* plant leaves. It is widely known for its strength, durability and versatility. Sisal plants yield approximately 200 $\pm$ 250 plant leaves, with each leaf containing 1000 $\pm$ 1200 fibre bundles made up of 4% fibre, 0.75% cuticle, 87.25% water and 8% dry matter [18]. It has various textile applications, both in its natural form and when blended with other fibres. Some textile applications of sisal fibre are ropes and twines, carpets and rugs, upholstery, geo-textiles, agro-textiles and fashion and accessories.

Lotus fibre—Lotus fibre (Fig. 2d) is a type of new-generation natural fibre. These fibres are obtained from the secondary wall of the lotus stem, located between the xylem tracheas. During the blooming time, a large number of lotus stems remain in the pond to rot. The leftovers or waste could provide a plentiful supply of natural cellulose fibre resources for use in the textile paper, healthcare and engineering sectors. Some of the properties of lotus fabric are similar to silk fabric such as light weight, flexibility, elasticity, delicate, very comfortable, high germicidal resistance, free of folds and high moisture absorbency. Currently, lotus fibres are widely employed in the production of luxury clothing, and garments made from lotus fibres are becoming increasingly popular due to their environmental friendliness and comfort [19].

Hemp fibre—Hemp fibre derived from hemp plants plays a very important role in the textile industry. Hemp fibre has high strength and is superior to other natural fibres in both cultivation and application. The cultivation of hemp fibre requires fewer chemicals and less water compared to cotton. Thicker or coarser hemp fibres and yarn are used to make cables, rope, sacking and robust tarpaulins, whilst fine hemp fibres are utilised in interior design and clothing. Tapestries, caps, shawls and towels are also made from it.

Milk protein fibre—Milk protein fibre is also known as milk silk or casein fibre. It is made from casein proteins found in milk. These fibres are smooth, soft, eco-friendly and biodegradable. These protein fibres have several applications such as clothing, lingerie and sportswear.

3 Properties of New-Gen Fibres

3.1 Physical Properties

New-gen fibres not only have unique physical and mechanical properties but also encompass a wider range of innovative materials. These specific properties can vary from fibre to fibre which depends on their source of origin. Some of these properties of new-gen fibres are as follows:

(1) **Tensile strength**: Many new-gen fibres can withstand pulling or force of stretching without any breakage or damage to the fibre, this is known as strength or tensile strength of the fibre. This property of these fibres makes them suitable for those applications, where strength, durability and resiliency of the fibres are required.
(2) **Lightweight**: New-gen fibres are light in weight and this property makes them suitable for those applications where weight reduction is desired. Due to their lightweight, these fibres are extensively used in different fields such

as the aerospace industry, automotive, composites and manufacturing of sports equipment.

(3) **Flexibility**: The new-gen textile fibres also confirm an excellent amount of flexibility, which allows these fibres to bend in different shapes without losing their structure. Flexibility is a very important or advantageous property in various applications, where the flexibility of the fibre is required for example wearable electronic textiles and textiles needed for free body movement.

(4) **Thermal properties**: Some fibres show exceptional thermal insulation properties such as high heat resistance and low thermal conductivity. These types of textiles are crucial where heat insulation is required.

(5) **Moisture management**: The moisture management property of the new-gen fibres is advantageous for outdoor clothing, sportswear and other moisture-sensitive applications. Some new-gen fibres exhibit high moisture absorbency which enables efficient absorption and transport of moisture away from the skin or fabric itself.

(6) **Thermal conductivity**: Some new-gen fibres transmit electricity or function as sensors due to their conductive and semi-conductive nature. Wearable electronics, electronic sensors and smart textiles are a few examples of the applications of these fibres.

(7) **Biodegradability**: Biodegradable and compostable natures of these fibres make them a sustainable and eco-friendly choice in apparel and other textile applications.

3.2 Chemical Properties

New-gen fibres exhibit unique chemical properties and these properties greatly depend on the type of fibre. Some of the chemical properties associated with new-gen fibres are as follows:

(1) **Chemical resistance**: Several new-gen fibres have resistance to acids, alkalis, solvents and oils. The chemical resistance of these fibres helps them to maintain their integrity and perform well when they are exposed to damaging substances.

(2) **Dyeability**: The dyeability of the new-gen fibre differs from fibre to fibre. Some fibres readily absorb dyes due to their inherent dyeability property whereas other fibres require chemical treatments to increase their dyeability to achieve desirable colours.

(3) **Flame resistance**: New-gen fibres also have flame resistance properties which means these fibres have excellent resistance to ignition and self-extinguish capabilities. This property is required in the areas where we need fire safety such as protective clothing for firefighters.

(4) **Antimicrobial properties**: Some new-gen fibres naturally have antimicrobial properties meaning they can inhibit the growth of bacteria, fungi and other microorganisms. These textile fibres are good for sportswear and healthcare

applications due to their antimicrobial properties, which prevents the formation of odour and maintain the freshness in the textiles.

(5) **UV resistance**: This property of fibres is required for those textiles where sun protection is required. These protect the wearer from harmful UV light. Outdoor clothing and awning are examples where sun protection is needed.

(6) **Eco-friendly properties**: Some of the new-gen fibres may be biodegradable or compostable, which leads to the sustainability of the textile industry. Additionally, few are made up of recycled or renewable resources which reduces the harmful impact on the environment.

3.3 Bio-degradability of New-Gen Fibres

New-gen fibres have the property of biodegradation which means these can be easily broken down by microorganisms. The biodegradability of new-gen fibres shows that these are eco-friendly and sustainable. Some biodegradability characteristics of these fibres are as follows:

1. **Biodegradability**: Biodegradation means the breakdown of organic matter by microorganisms. New-gen fibres can be decomposed and broken down into simple, no-toxic compounds by the microorganisms in natural environmental conditions which reduces landfill waste and adverse impact on nature.
2. **Compostability**: New-gen fibres have biodegradability as well as compostability. This means these fibres undergo control degradation in composting, where they break down into nutrient-rich organic matter that can be used for soil health improvement.

4 Spinnability of the Fibres

Spinnability means the ability of fibre to convert into yarn through the spinning process. The new-gen fibres have this property of spinnability they can be easily converted into yarn due to their inherent crimps and cohesiveness. The spinnability of fibres depends on various factors such as physical, and chemical properties, fibre structure and fibre processing conditions. Here are some factors on which fibre spinnability depends

1. **Fibre length and fineness**: The length of fibre is one of the most important factors which play a significant role in spinning. The long length of new-gen fibre provides better spinnability as compared to short-length fibres because these can be aligned and twisted easily at the time of the spinning process. Fine fibres require blending or supplementary techniques to improve their spinnability.
2. **Strength and elasticity**: Successful spinning of fibres requires good strength and elasticity. Natural fibres can withstand mechanical forces and tension that are exerted during the spinning process without any fibre breakage. The elasticity

of these new-gen fibres helps them to recover in their original shape and lower the breakage at the time of twisting and drafting.

3. **Fibre cohesiveness**: The ability of new-gen fibres to stick to each other and develop a cohesive structure is required for successful spinning. Fibre with cohesiveness properties can spin well and form a continuous, smooth and stable yarn. Sometimes surface modification and treatment are required to build cohesiveness.
4. **Surface properties of the fibre**: Smoothness, roughness, hydrophilic and hydrophilic nature of fibres can affect the spinning of the fibres. The alignment of fibres at the time of spinning can be improved by the smoothness and uniform surface of the fibres. The surface of rough and coarser fibres can be improved by the surface treatment or coating of the fibres.
5. **Processing techniques of fibre**: The spinning technique plays a significant role in the spinning of new-gen fibres. With various fibre spinning, air jet spinning or open-end spinning, ring spinning may have distinct effects. Few new-gen fibres needed modification in the traditional spinning process or to optimise spinnability development of specialised spinning technique.
6. **Blending of fibres**: Natural or synthetic fibres can be blended with the new-gen fibres which improves spinnability. Cohesiveness, alignment of fibres and overall processing characteristics can be improved through blending and it leads to better yarn formation.
7. **Processing condition**: Processing conditions may vary according to the specific fibre properties and required yarn structure. Spinning parameters such as tension, twist, drafting and temperature must be optimised for new-gen fibres to achieve desired spinnability.

5 Sustainable Industrial Applications of New-Gen Fibres

New-gen textile fibres offer sustainable solutions for various industrial applications, contributing to a more environmentally friendly and socially responsible textile industry. People in both industrialised and developing countries are rapidly gravitating towards the usage of renewable resources to address wellness problems as well as environmental concerns. As a result of growing anticipates about natural products, manufacturers are looking for environmentally friendly, sustainable, and materials that are recyclable. However, in the case of textiles, the manufacturing of natural products from fibres that are renewable plays an important role. The main four usable principles of sustainability are economy, society, culture and environment. Considering the next generations while keeping all of the recyclables and excess materials in mind. To confront and resolve these difficulties, we require comprehensive and effective organisational structures, as well as resources to ensure long-term viability. The application of synthetic clothing harms the surroundings, which has an indirect impact on our society and economy. According to 'Environmental and Health Concerns', commodities must be eco-friendly and also support the economic conditions of a country [24]. The automobile sector is effectively using composites augmented with natural fibres to substitute elements including dashboard panels and cushions for seats that were formerly developed from fibreglass mats or polymeric foams [25]. Animal-based fibres, in addition to plant-based fibres, have emerged as feasible options for generating bio-degradable, biomedical and bio-resorbable materials of composite for bio-engineering and surgical uses. Sutures made of silk fibre have been employed in bio-medical implementations [26].

Some sustainable industrial applications of new-gen fibres:

1. **Sustainable apparel**: New-generation fibres, such as recycled polyester, organic cotton, and cellulose fibres that have been regenerated like lyocell and modal, are used in the production of sustainable apparel. These fibres reduce reliance on non-renewable resources; minimise energy consumption and lower greenhouse gas emissions compared to conventional fibres.
2. **Technical textiles**: New-generation fibres find applications in technical textiles, which are functional textiles used in industries such as automotive, construction, and health care. Fibres with properties like high strength, flame resistance, moisture-wicking and antibacterial characteristics are utilised to create sustainable technical textiles that meet specific performance requirements.
3. **Protective clothing**: Sustainable fibres with inherent properties like flame resistance, UV protection and chemical resistance are used in the production of sustainable protective clothing. These fibres ensure worker safety while minimising environmental impact through reduced use of hazardous chemicals and improved end-of-life options.
4. **Filtration and separation**: Most of the new-gen fibres are used for filtration and separation application due to their fine structure. These fibres enhance the filtration, increase surface area and have improved adsorption, which reduce waste generation and improves air and water quality.
5. **Geotextiles**: In civil engineering and building construction, geotextiles utilise new-gen textile fibres. These fibres offer tremendous strength, soil stabilisation, drainage, and erosion management and also lowering the requirement for conventional materials and having a smaller negative impact on the environment.
6. **Nonwoven**: Wipes, medical textiles and sanitary products are a few examples of nonwoven applications that make use of new-generation fibres. Nowadays, environmentally friendly alternatives like bamboo, hemp, or recycled polyester are used in place of conventional materials and minimise the impact of these items on the environment.
7. **Automotive textiles**: In automobile textiles, sustainable fibres, including natural fibres like hemp and kenaf, are used. With the advantages of being recyclable and biodegradable, these fibres provide lightweight substitutes to conventional materials that lighten the weight of vehicles and increase fuel efficiency.
8. **Sports and outdoor textiles**: Sportswear and outdoor fabrics made from sustainable materials use new-gen fibres. Performance clothing that is eco-friendly, comfortable for wearers, and stain-resistant is made possible by new-gen fibres having these qualities.

6 New Generation Textile Fibres and Environment Conservation

New-generation textile fibres play a significant role in promoting environmental conservation within the textile industry. Here are some ways in which these fibres contribute to environmental conservation:

1. **Use of renewable resources**: Most of the new-gen fibres used in textiles are made from sources that are renewable including bamboo, hemp and organic cotton. These fibres decrease the dependency on conventional resources like petroleum-based synthetic fibres, minimising environmental impact and promoting sustainable agriculture practices.
2. **Reduction of chemical usage**: New-gen fibres often require fewer chemical inputs during production compared to conventional fibres. For instance, organic cotton fibres are grown without the use of synthetic pesticides and fertilisers, leading to the reduction of water pollution and soil degradation. This helps to conserve ecosystems, protect biodiversity and safeguard the health of farmers and workers.
3. **Recycling and up-cycling**: Certain new-gen fibres, like recycled polyester and regenerated cellulose fibres, are produced from post-consumer or post-industrial waste materials. By recycling and up-cycling these materials, the textile industry reduces waste sent to landfills, conserves energy and resources required for virgin fibre production, and mitigates the environmental impact of textile waste.
4. **Energy and water efficiency**: Some new-gen fibres, such as lyocell and modal, are manufactured using closed-loop processes that recycle and reuse chemicals and solvents. These processes minimise water consumption and energy requirements, reducing the overall environmental footprint of fibre production.
5. **Biodegradability and compostability**: Some modern fibres, including hemp and bamboo, are inherently compostable and biodegradable. The capacity of these fibres to degrade at the final stage of their life cycle without releasing dangerous elements into the environment will result in an improved efficient and sustainable textile business.
6. **Reduced carbon footprint**: When compared to conventional fibres, the manufacture of new-gen fibres frequently results in fewer greenhouse gas emissions. As organic cotton cultivation uses less energy and doesn't use synthetic fertilisers, it often has a lower carbon footprint. This supports environmental sustainability and lessens the effects of climate change.
7. **Sustainable manufacturing**: New-gen fibres for fabrics are frequently generated with sustainable production techniques that are eco-friendly to the environment. This includes eco-friendly dyes, finishes, water management, energy use and waste minimisation techniques. The whole textile sector benefits from these environmentally friendly practices.

7 Problems and Future Prospects

Although they present the textile industry with exciting potential, new-generation textile fibres also have several obstacles to overcome. The followings are some difficulties and prospects for the future relating to new-generation textile fibres:

(1) **Scalability and cost**: To meet industry demand and keep costs competitive, one of the major difficulties is to increase the production of new-gen fibres. To increase the marketability of these fibres, efficient and affordable manufacturing procedures must be developed.
(2) **Performance and durability**: To compete with conventional fibres, new-gen fibres frequently need to fulfil particular performance requirements. These fibres must exhibit sufficient strength, durability and other functional properties for usage in a range of applications.
(3) **Transparency and traceability in the supply chain**: It is critical to establish transparent and traceable supply chains for new-gen fibres to validate their sustainability claims, such as organic or recycled content. Ensuring dependable certification processes and traceability methods would aid in the development of consumer and company trust.
(4) **Recyclability and the circular economy**: Although many new-generation fibres can be recycled, it is critical to building adequate recycling infrastructure and processes to optimise their circularity. It will be critical to developing technology and systems to efficiently collect, separate, and recycle these fibres to minimise wastage and promote a circular economy.
(5) **Consumer awareness and adoption**: Growing consumer knowledge and comprehension of the benefits of next-generation fibres is critical to boosting demand and acceptance. Educating customers about the environmental and social benefits of these fibres can increase their selection and market acceptance.
(6) **Research and innovation**: Further efforts in research and development are required to improve the properties and performance of next-generation fibres. Investing in innovation, such as developing novel spinning techniques, improving fibre characteristics, and investigating new fibre sources, will open up new possibilities and expand the potential for the use of these fibres.
(7) **Collaboration and partnership**: Coordination among fibre manufacturers, textile industry stakeholders, research institutions, and policymakers is critical for tackling difficulties and propelling the future of next-generation textile fibres. Collaborative activities can aid in the streamlining of processes, the sharing of knowledge, and the establishment of shared sustainable practices.
(8) **Standardisation and the regulatory framework**: Creating and implementing regulatory frameworks and standards tailored to new-generation fibres will provide direction, assure uniformity and promote responsible practices throughout the industry. The environmental labelling standards, environmental impact studies and labelling requirements can all serve to establish confidence and facilitate informed decision-making.

8 Conclusions

New-generation textile fibres have enormous promise for transforming the textile industry into one that is more environmentally conscious and sustainable. These fibres provide several advantages, including less reliance on non-renewable resources, a lower carbon footprint, better waste management and increased performance qualities. They enable the production of environmentally friendly clothes, technical textiles, protective gear and a variety of other applications. However, scalability, cost, performance, recycling infrastructure and consumer awareness remain barriers to the widespread adoption of next-generation textile fibres. Overcoming these difficulties will necessitate the collaboration of industry players, policymakers, researchers, and consumers.

Despite these obstacles, the future of next-generation textile fibres seems bright. Improvements in fibre quality, production efficiency and sustainability will be driven by advances in manufacturing methods, research and innovation. Increased customer demand for environmentally friendly and sustainable products will accelerate the acceptance and development of these fibres. Finally, the success of next-generation textile fibres will be determined by the development of a more sustainable and responsible textile sector that reduces its environmental effect, conserves resources and promotes social well-being. The textile industry can contribute to a greener future and build a more sustainable and circular economy by embracing these fibres and encouraging sustainable practises across the supply chain.

References

1. Akira N (2000) Fiber science and technology
2. Ganster J, Fink HP (2009) The structure of man-made cellulosic fibres. In: Handbook of textile fibre structure. Woodhead Publishing Series in Textiles, Woodhead Publishing, Sawston
3. Nair V, Khosla P, Ramachandran M (2016) Review on mechanical properties of various natural fibres reinforced composites. Res J Pharm Biol Chem Sci
4. Velde KV, Kiekens P (2001) Compos Struct 54:355
5. Murphy WS (2003) Preparation of textile fiber. Abishek Publication
6. Sahu P, Gupta MK (2017) Sisal (Agave sisalana) fibre and its polymer-based composites: a review on current developments
7. Yousif BF, Shalwan A, Chin CW, Ming KC (2012) Flexural properties of treated and untreated kenaf/epoxy composites. Mater Des 40:378–385
8. Reddy N, Yang YQ (2005) Bio fibres from agricultural by products for industrial applications. Trends Biotechnol
9. Yang X, Huang L, Cheng L, Yu J (2012) Studies of moisture absorption and release behaviour of Akund fiber. Adv Mech Eng
10. Rowell RM (2008) Natural fibres: types and properties. Woodhead Publishing Limited
11. Shuvo II (2020) Fibre attributes and mapping the cultivar influence of different industrial cellulosic crops (cotton, hemp, flax, and canola) on textile properties. Bioresources Bioprocess 7(1). https://doi.org/10.1186/s40643-020-00339-1
12. Zakikhani P, Zahari R, Sultan MTH, Majid DL (2014) Extraction and preparation of bamboo fibre-reinforced composites. Mach Des 63:820–828

13. Venkateshwaran N, Elayaperumal A (2010) Banana fiber reinforced polymer composites—a review. J Reinf Plast Compos 29:2387–2396
14. Gupta US, Dhamarikar M, Dhakar A, Tiwari S, Namdeo R (2020) Study on the effects of fibre volume percentage on banana—reinforced epoxy composite by finite-element method. Adv Compo Hybrid Mater
15. Kumar MD (2008) A study of short areca fiber reinforced PF composites. In: Proceedings of the world congress on engineering, WCE, London
16. Shinoj S, Visvanathan R, Panigrahi S, Kochubabu M (2011) Oil palm fiber (OPF) and its composites: a review. Ind Crops Prod 33:7–22
17. Mishra S, Mohanty AK, Drzal LT, Misra M, Hinrichsen G (2004) A review on pineapple leaf fibers, sisal fibers and their biocomposites. Macromol Mater Eng 289:955–974
18. Li Y, Mai YW, Ye L (2000) Sisal fibre and its composites: a review of recent developments. Compos Sci Technol 60:2037–2055
19. Yousif BF, Shalwan A, Chin CW, Ming KC (2012) Flexural properties of treated and untreated kenaf/epoxy composites. Mater Des
20. Go Green Products (n.d.) https://gogreenproducts.org/natural-fiber/
21. Nesher A (2020) Plant-based fibre: fashion businesses that address microplastic pollution. The Vegan review. https://theveganreview.com/plant-based-fibre-fashion-businesses-that-address-microplastic-pollution/
22. Bhoyar RK, Kothe RP, Gillarkar S, Chacherkar S, Barve R, Mude P (2020) Studies on mechanical behavior of sisal fiber and human hair hybrid sandwich composites. Int Res J Eng Technol (IRJET) 7(5)
23. Textilecoach (2021) Lotus fibre. https://www.textilecoach.net/post/lotus-fibre
24. Ali MA, Sarw MI (2010) M.Sc thesis "sustainable and environmental friendly fibres in textile fashion—a study of organic cotton and bamboo fibres". Applied Textile Management, University of Borås
25. Monteiro SN, Lopes FPD, Ferreira AS, Nascimento DCO (2009) Natural- fibre polymer matrix composites: cheaper, tougher, and environmentally friendly. JOM 61(1)
26. Cheung HY, Ho MP, Lau KT, Cardona F, Hui D (2009) Natural fibre-reinforced composites for bioengineering and environmental engineering applications. Composites Part B 40:655–663

Role of Chemicals in Textile Processing and Its Alternatives

M. Pavan, Lata Samant, Surabhi Mahajan, and Manpreet Kaur

Abstract Chemical substances and additives have a crucial function in the manufacturing of textiles, starting from the elimination of impurities from fibers to providing the ultimate touch to the fabric. The textile industry predominantly utilizes chemicals throughout multiple stages of processing, including sizing, de-sizing, yarn processing, and fabric and garment production. Despite being a significant revenue generator, this industry poses grave environmental risks and is a major contributor to pollution worldwide. The gush in the use of synthetic chemicals and their associated vulnerabilities has led the textile sector to implement sustainable policies. The exigency of chemicals in the fashion industry is under immense force owing to a burgeoning population and swelling consumer demand for fashionable fabrics, leading to the production of more products than ever before. The present chapter provides an in-depth exploration of the various chemical substances utilized throughout various stages of textile processing. Moreover, this chapter sheds light on the detrimental effects of conventional chemicals employed in textile processing procedures, concurrently proposing environmentally friendly and sustainable substitutes.

Keywords Chemicals · Textile processing · Environmental risks · Sustainable substitutes

M. Pavan (✉) · S. Mahajan
Department of Apparel and Textile Science, Punjab Agricultural University, Ludhiana, Punjab 141004, India
e-mail: pavanmanu1212@gmail.com

L. Samant
Department of Clothing and Textiles, GBPUAT, Pantnagar, Uttarakhand 263145, India

M. Kaur
Department of Textile and Apparel Designing, University of Agricultural Sciences, Dharwad, Karnataka 580005, India

Sadhna et al. (eds.), *Climate Action Through Eco-Friendly Textiles*, SDGs and Textiles,
https://doi.org/10.1007/978-981-99-9856-2_5

1 Introduction

In the nineteenth century, the apparel industry played a pivotal part in driving the global transition toward industrialization. However, as a consequence of this rapid industrial growth, the issue of freshwater pollution caused by industrial activities has become increasingly prominent. Each new mechanical innovation in the Industrial Revolution likely had to spur additional chemical advancements in the textile industries. Much like how handlooms struggled to keep up with demand after the advent of the spinning jenny, it's probable that the processes of bleaching, dyeing, and other chemical treatments encountered challenges in meeting the growing demands as textile innovations boosted the production of woven goods. Back then, as well as today, the public places a significant value on the appearance of fabrics. Any imperfections resulting from chemical processes would undoubtedly have harmed the sales of these materials [1], highlighting how chemicals have become an indispensable part of the wet processing of textiles.

Textile wet-processing refers to the crucial stage in textile production where the textile substrate undergoes treatment with various chemicals, colorants, and auxiliary substances within an aqueous medium, typically water. The conventional wet-processing of textiles can be divided into three essential stages:

(a) Preparation stage: This stage includes operations such as sizing, scouring, desizing, mercerization, and bleaching. These processes are crucial for preparing the textile substrate for further treatment and applications.
(b) Coloration stage: In this phase, the textile materials undergo dyeing and printing processes to achieve the desired colors, patterns, and designs. It is a crucial step for adding aesthetic appeal to textiles.
(c) Finishing and laundry: The final stage involves finishing treatments to improve the texture, appearance, and functionality of the textiles. Additionally, laundering processes may be employed to enhance softness and remove any residual chemicals or impurities.

These three stages collectively contribute to the transformation of raw textile materials into the finished textile products that we use in our everyday lives.

In textile wet-processing, it is common practice to use chemicals and auxiliaries that are often more toxic, less biodegradable, and potentially hazardous to our environment. Unfortunately, this sector stands out as one of the most significant contributors to pollution and effluent discharge in the fashion industry, primarily due to excessive water consumption, chemical dyes, as well as auxiliary substances employed in various processing operations [2–4].

According to a report by Greenpeace International in 2011, it was found that around 25% of the total chemicals produced globally are specifically employed within the textile industry. In textile wet-processing alone, there are approximately 2000 different chemicals in use. Throughout the various stages of textile wet-processing, it has been observed that a significant number of these chemicals exhibit volatility or solubility in water, while others tend to persist within the fabric. As a result, they pose potential risks to humans, plants, animals, and aquatic algae [5].

Moreover, effluent discharges comprise a complex mixture of various compound contaminants, including colorants leached from the fabric, high organic content, and inorganic compounds like sodium chloride (NaCl), sodium hydroxide (NaOH), hydrochloric acid (HCl), and laundry detergents [6, 7]. Table 1 lists various toxic chemicals employed in the wet processing of textiles. Wastewater generated during different textile processing operations contains substantial pollutant levels that can be highly detrimental to the ecosystem if released without appropriate treatment.

Substances that have the potential to harm the environment often exhibit the following characteristics:

- Persistence: These substances do not easily decompose in the surroundings.
- Bioaccumulation: These substances can build up in the tissues of beings as they move up the food chain.
- Toxicity: These substances have the potential to be toxic or detrimental to surrounding beings.

Hence, any chemical substances exhibiting these characteristics are referred to as *PBT* (*Persistent; Bioaccumulative; Toxic*) substances. Organic materials displaying these particular characteristics are denoted as *Persistent Organic Pollutant* (*POP*) substances. Despite undergoing initial dispersion in significant quantities of water or air, it is notable that pollutants can persist in the receiving region for a considerable duration. This persistence enables their transportation across wide distances,

Table 1 List of toxic chemicals employed in various textile wet processes [8]

Name of the chemical	Applications
Acetic acid	Neutralization
Azo dyes	Textiles dyeing process
Surfactants (Cationic)	Dyeing, and finishing
Chlorine-based bleaches	Textile bleaching and disinfection
Copper (elemental, non-complexed form)	Dyeing of cotton, and polyamide (Blue and black dye)
Cyanide	Anticaking additive
Formaldehyde	Textile dyeing and finishing
Surfactants (Nonionic)	Laundry detergents employed in the preparation and dyeing of textiles
Perfluorinated chemicals (PFCs)	Water and stain-resistant finishing
Nonylphenol ethoxylates (NPEs)	Fabric processing, and detergents
Peroxide bleaches	Bleaching
Phthalates	Textile printing (inks), and coating
Sodium chloride (NaCl)	Dyeing of cotton materials (exhausting agents)
Sodium hydroxide (NaOH)	Scouring, mercerization
Sodium sulfate (Na_2SO_4)	Dyeing of cotton materials (exhausting agents)
Tributyltin oxide (TBTO)	Biocides in the textile industry

their accumulation in sediments and living organisms, and their potential to inflict substantial harm, even at seemingly low concentrations.

The catalogs of banned substances are constantly evolving as experts and medical professionals provide more data, leading to a better understanding of how chemicals affect the health of human beings and the ecosystem. The elements found within the *Restricted Substances List* (*RSL*) are primarily influenced by global regulations that govern the application of chemicals in the processing of clothing and other materials. The European Union (EU) has implemented the 'Regulation Concerning the Registration, Evaluation, Authorization, and Restriction of Chemicals' as a measure to safeguard human health and the environment from potential risks associated with chemical substances [9, 10].

The comprehensive inventory of substances, materials, and chemicals known as the *Restricted Substances List* (*RSL*) is issued by the *American Apparel and Footwear Association* (*AAFA*). This list specifically pertains to apparel, footwear, and home textile products, and it outlines substances that are either prohibited or subject to limitations within these industries. The aforementioned restrictions have been implemented as a direct consequence of prevailing regulations or legislation. The *Restricted Substances List* (*RSL*) encompasses a diverse array of substances, not limited to disperse dyes with specific properties, but includes substances like arylamines, pesticides, solvents, asbestos, dioxins and furans, Greenhouse Gas Emissions (GHGs), flame retardants, metals, specific fluorinated compounds, phthalates, organotin compounds, and a variety of other compounds. The presented compilation of information serves as a highly valuable and indispensable resource for guaranteeing adherence to regulatory standards within the industry [1, 10].

Nonylphenol Ethoxylate (*NPE*) is a substance that is widely employed in the textile industry for various applications such as cleaning, dyeing, and rinsing. When released into the environment, *NPE*s transform, leading to the formation of various substances, including nonylphenol. This specific compound tends to accumulate in the bodies of fish, disrupting their hormone levels and detrimentally affecting their fertility, sexual growth, and development. Although the use of *NPE* in the production of textiles has been prohibited in Europe for a considerable period, it continues to be released into aquatic environments due to the washing of imported textiles. Recognizing the unacceptable environmental risks it poses, EU members have collectively decided to extensively ban *NPEs* [10, 11].

The industry is adopting several strategies to address these concerns. These approaches include: (a) Avoidance: Prioritizing the prevention of unsafe chemicals in textile processing to minimize their use from the outset; (b) Reduction: Efforts to reduce the overall usage of such chemicals wherever possible; (c) Substitution: Replacing hazardous chemical processes with safer alternatives to eliminate potential risks. Driven by factors like growing demand for nature-friendly processes and stricter regulations on polluting innovations, there is a rising trend in the utilization of biotechnology in the textile manufacturing sector [12]. This shift toward safer and more sustainable practices is helping to mitigate the environmental and health impacts associated with traditional textile processing.

2 Overview of Chemicals Utilized in the Textile Manufacturing Sector

The textile industry relies on a diverse range of chemicals, and auxiliary substances to effectively process textile substrates and produce finished textile products. Each chemical serves a distinct purpose within the various stages of processing. For instance, caustic soda (sodium hydroxide) is used in the scouring process to eliminate impurities, while in mercerization, it enhances the luster, strength, and dye affinity of cotton fabric. Liquid chlorine, on the other hand, finds application in processes such as bleaching, disinfection, oxidation, and the production of other chemicals within the textile industry. Other significant chemicals employed in textile wet-processing include soda ash (sodium carbonate), hydrogen peroxide, hydrogen chloride, and acetic acid, as well as various dyes and pigments. These chemicals play crucial roles in achieving the desired characteristics and qualities of textile products.

Figure 1 illustrates the production trends of major chemicals, dyes, and pigments in India, spanning from 2016–17 to 2020–21 (These chemicals are primarily used in textile industries or laboratories, where they are involved either directly or indirectly in various textile processes. Their applications range from fabric preparation and treatment in textile factories to laboratory testing and research related to textile materials and products). Notably, caustic soda, with a production volume of 2964.12 million tons, emerged as one of the most produced chemicals in the year 2020–21, closely followed by soda ash, which accounted for 2638.08 million tons. Liquid chlorine, hydrogen peroxide, and acetic acid production reached 2174.26 million tons, 139.9 million tons, and 154.76 million tons, respectively, during the same year. Meanwhile, dye production stood at 236.46 million tons, and pigment production amounted to 90.5 million tons in 2020–21. Despite the marginal decline in the production trends of these chemicals in the year 2020–21, it's noteworthy that the demand for these chemicals continues to rise at an alarming rate [13, 14]. This indicates the enduring importance of these substances in various industries, including textiles, and suggests a growing need for them in the foreseeable future.

3 Preparatory Processes, Chemicals Used, and Their Alternatives

3.1 Sizing

Sizing is the process of applying a sizing agent to the warp yarns to resist yarn wear during weaving operations from friction, stretching, and bending. Sizing significantly reduces yarn hairiness and improves yarn strength and abrasion resistance. Commonly, starch, carboxymethyl cellulose (CMC), animal glue, acrylic copolymers, Polyvinyl Alcohol (PVA), water-soluble polyesters, etc. are employed

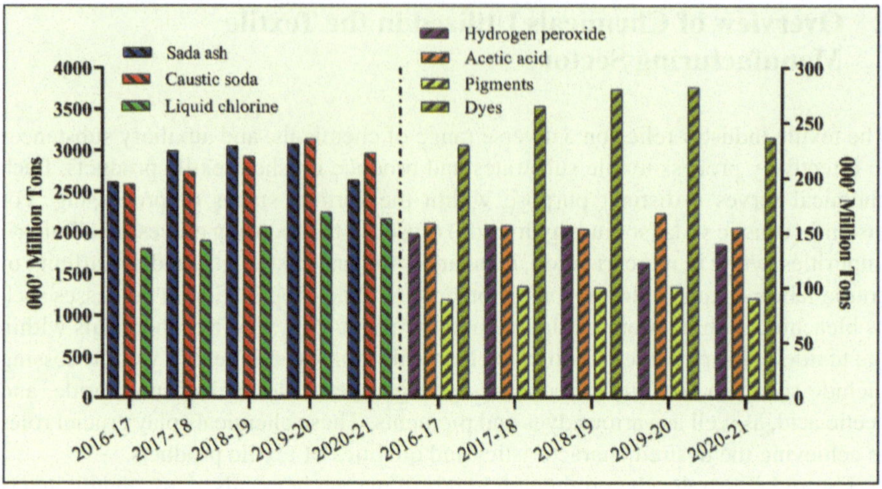

Fig. 1 Production scenario of major chemicals for textile processing from the years 2016–17 to 2020–21 in India [13, 14]

for sizing. These sizing agents accompany various auxiliaries like softeners, preservatives, surfactants, and waxes (paraffin wax, bee wax, and synthetic wax) [15].

3.2 Desizing

The elimination of sizing materials from woven textiles is referred to as desizing. The presence of these sizes may impede the dyeing, printing, and finishing processes. Desizing is a prerequisite to other wet processes such as mercerizing, bleaching, printing, or dyeing. The desizing techniques and chemicals employed are determined by the kind of sizing agent utilized. There are two primary desizing approaches: hydrolytic methods (acid steeping, alkali steeping, enzymatic steeping, and rot steeping) and oxidative methods (chlorine, chlorite, bromide, peroxy compounds, and ammonium persulfate desizing. The desizing agents commonly used in these methods include sodium hydroxide, hydrochloric acid, bleaching powder or sodium hypochlorite, sodium chlorite, sodium persulfate or hydrogen peroxide, and others [16, 17].

3.3 *Scouring*

During the scouring phase, the fabrics undergo a cleaning process to eliminate impurities such as gums, lubricants, waxes, and grime. This is carried out in an alkaline setting and at high temperatures, which transforms these contaminants into water-soluble substances. To boost the efficacy of the process, scouring baths often contain emulsifiers, wetting agents, surfactants, reducing agents, and sequestrants. So, after scouring, the wastewater has a lot of alkaline substances with high levels of Biochemical Oxygen Demand (BOD), Chemical Oxygen Demand (COD), and Total Dissolved Solids (TDS) [4, 18].

3.4 *Bleaching*

The bleaching process is essential for removing unwanted natural color elements, giving textile materials a whiter appearance. Achieving optimal bleaching results depends on the substrate, and it involves the use of various chemicals and application techniques. However, it's equally important to eliminate any excess or residual bleaching agents from textile goods before proceeding with further wet processing. There are two main types of chemical processes used for bleaching: oxidative and reductive. The choice between these processes depends on the substrate and the nature of the color impurities. Oxidative bleaches include bleaching powder, sodium hypochlorite (NaClO), hydrogen peroxide (H_2O_2), sodium persulfate ($Na_2S_2O_8$), sodium percarbonate ($Na_2H_3CO_6$), sodium chlorite ($NaClO_2$), peracetic acid, and sodium perborate. On the other hand, reductive bleaches encompass sodium hydrosulfite ($Na_2S_2O_4$), sodium thiosulfate ($Na_2S_2O_3 \cdot xH_2O$), and derivatives of sulfinic acid. Each of these agents performs a decisive part in achieving the desired bleaching outcomes in textile processing [19].

3.5 *Mercerization*

Mercerization, a crucial stage in textile pretreatment, involves the utilization of sodium hydroxide (commonly known as caustic soda, NaOH) to enhance the luster and dyeability of cotton yarns or textiles. The application of this treatment has been observed to result in notable improvements in both the aesthetic and functional aspects of cotton material. Consequently, the treated cotton exhibits enhanced suitability for a wide range of textile applications [20, 21].

3.6 Dyeing and Printing

Dyeing is a widely employed technique in the textile industry that aims to improve the aesthetic appeal of fabrics by the application of diverse hues and tones. The application of this coloring process is versatile, as it can be implemented at various stages within the production of textile materials. These stages encompass the treatment of fibers, yarns, fabrics, and even finished textile products such as clothing and accessories. Textile dyes are chemical substances that possess the ability to selectively absorb and emit light at specific wavelengths, thereby influencing the perception of color by the human visual system. The process of dyeing entails a complex interaction between dyes and textile fibers, whereby the dyes effectively enter the structure of the fibers. Any dye that doesn't bond with the fibers during the coloring process endures in the dye bath and is typically released as effluent. As a result, it becomes necessary to treat these effluents to prevent potential environmental hazards associated with the release of dye residues into the environment. Proper effluent treatment is essential to mitigate the impact of dyeing processes on the ecosystem [8, 22].

Using sodium sulfate and sodium chloride as exhausting agents used during direct dying of cotton goods is a big problem, especially in areas where the natural flow of receiving streams is much lower than the capacity of POTW (Primary Oxygenation Treatment Water). Regrettably, there is currently no practical method for removing these salts from wastewater. Consequently, the primary approach has been to dilute the hazardous discharge. Nevertheless, these problems can be mitigated by employing low-salt-based reactive dyes along with applying padding methods. An additional concern lies in the presence of copper in numerous black and blue dyes, with *free* copper representing an instant lethal element. Consequently, there has been a collective and dedicated attempt to screen and improve copper-free dyes, reducing the environmental impact of these substances in textile processes.

Textile printing is the process of applying color or designs onto fabric in a precise and controlled manner. This technique encompasses the application of decorative patterns to the fabric through various methods like rotary screens, rollers, or flat screens [23]. One key contributor to emissions in print adhesives is the solvent used, which can be organic, aqueous, or a blend of both. The proportion of organic solvent to water in printing pastes can vary widely, with ratios ranging from 0 to 60% based on fabric weight. There is no standard or fixed proportion for the organic solvent-to-water ratio in printing pastes, as it depends on the specific requirements of the printing process and the type of fabric being used [8, 24].

High-density metals belong to a group of metallic elements characterized by atomic numbers between 22 to 34 and 40 to 52. This category encompasses elements found in both the actinide and lanthanide series, and they demonstrate a density that surpasses that of water by four to five times. In textile processes, heavy metals can originate from various sources, including raw fibers, water, colorants (which may contain heavy metals as components), additives, residual chemicals from finishing processes, and plumbing fixtures utilized in dyeing and finishing areas [5]. Table 2

Table 2 Health risks linked to the use of heavy metal and metalloid in textile sector [5, 25]

Heavy metal/ metalloid	Health hazards
Arsenic	Kurtosis, hyper-pigmentation, black foot disease, skin, lung, kidney and bladder cancer
Cadmium	Ailments of the lungs, kidneys, and bone system
Chromium	Lung and skin disorders, cancer of the respiratory tract on inhalation, and ulceration of skin
Lead	Damage brain, nervous system, and kidneys. Also triggers mild instances of anemia, insomnia, loss of appetite, restlessness, and gastrointestinal illness
Mercury	Brain damage, tremors, memory loss, and kidney damage
Selenium	Skin rashes, nail and hair loss, and gastrointestinal issues
Nickel	Respiratory problems, lung cancer, and skin allergies
Copper	Liver and kidney damage, central nervous system disorders, and gastrointestinal distress

provides an overview of some of these heavy metals along with their potential health hazards in the textile industry.

3.7 *Finishing and Laundry*

The water-repellent coating is rendered onto the textile materials to make them resistant against water, stain, and oil. This process typically employs polyfluoroalkyl substances for Durable Water-Resistant (DWR) finished products. Polyfluoroalkyl substances can be categorized based on the length of their perfluoroalkyl moiety chain, falling into two groups: short- and long-chain variants. Where, long-chain PFA molecules are characterized by an alkyl chain having six or more carbon atoms in perfluoro-sulfonic acids (PFASs) molecule (C_nF_{2n+1} SO_3H, where n is greater than or equal to 6) and seven or more carbon atoms in perfluoro-carboxylic acid (PFCA) molecules (C_nF_{2n+1} COOH, where n is greater than or equal to 7).

However, due to mounting concerns regarding the detrimental impacts of employing long-chain poly-fluoroalkyl substances over human beings and the ecosystem, efforts are underway to transition towards short-chain polyfluoroalkyl substances or non-fluorinated substitutes from long-chain polyfluoroalkyl substances. Non-fluorinated durable water-repellent options that are now available, include silicones, paraffin waxes, as well as other molecules like dendrimers and inorganic nanoparticles [26, 27].

PFCs (Perfluorinated Chemicals) are synthetic compounds that are exclusively generated through anthropogenic means, meaning they do not arise naturally in the environment. In addition to being largely insoluble in water and oils, they also exhibit exceptional resistance to thermal, biological, and chemical deterioration. The extensive utilization of these compounds can be attributed to their distinctive characteristics, which render them highly effective as coatings for textiles and papers, as well as efficient solvents and surfactants in various industrial settings. Moreover, they serve as essential components in cosmetics, fire-fighting foams, and plastics, while also finding application as lubricants in high-temperature environments. These PFCs are classified as follows: PFASs (Perfluoroalkyl Sulfonates; mostly known as Perfluorooctane Sulfonic Acid); PFCAs (Perfluorinated Carboxylic Acids; mostly known as Perfluorooctanoic Acid); Fluoro-polymers (mostly known as PTFE (Poly-Tetra-Fluoro-Ethylene)), which is widely known as Teflon and is the basis for non-stick cookware and waterproof fabric); and Fluorotelomer alcohols [28, 29].

The present market for low-maintenance and durable press textile finishes is ruled by 2-Imidazolidinone analogs, precisely Dimethylolethylene urea (DMEU) byproducts. The compound that has gained significant popularity for cross-linking cellulosic fibers is DMDHEU (N, N′-1,3-dimethylol-4, 5-dihydroxyethylene urea). This particular compound is commonly used in conjunction with magnesium chloride ($MgCl_2$), as an acidic catalyst for commencing the bonding process. The cross-linking treatment is typically carried out using a pad-dry-cure method. Since the 1970s, there has been a notable increase in the utilization of DMDHEU compared to melamine formaldehyde and urea formaldehyde (dimethylol urea) resins. This shift can be attributed to the desire to achieve enhanced sustaining properties, driven by efforts to reduce formaldehyde releases. In addition to the potential direct release of formaldehyde, which has been linked to carcinogenicity, it is also a recognized skin irritant. The presence of these linking chemicals over clothing that come into contact with the skin may pose potential risks of dermatitis. Chemical manufacturers are currently involved in a gradual transition toward the utilization of ultra-low formaldehyde-based bonding compounds. Furthermore, they are actively conducting investigations into various zero-formaldehyde-based cross-linking chemicals specifically designed for application on cellulose-based textiles. The utilization of glyoxal-urea and polycarboxylic acid-based cross-linking compounds has been observed as an effective substitute for formaldehyde-linked cross-linkers. This approach aims to address concerns regarding the hovering health hazards linked with formaldehyde exposure [30].

Phthalates, often referred to by the term plasticizers, perform diverse functions in textile coating and printing procedures, as well as in the manufacturing of synthetic rubber, leather, and selected dyes. One of their primary applications is softening PVC (Polyvinyl Chloride). Nevertheless, it is imperative to acknowledge the prevailing apprehensions of the potential toxicity associated with specific phthalates, particularly DEHP (bis (2-ethylhexyl) phthalate), which has been identified as having reprotoxic properties. This indicates that DEHP has negative effects on human embryos and gametes, which hinder the development of mammalian embryos and result in lower fertilization rates. In Europe, phthalates like Dibutyl Phthalate (DBP) and DEHP are categorized as 'hazardous to reproduction' and their usage is subject to restrictions. These concerns highlight the need for careful consideration and regulation of phthalate use in various industries, including textiles, to safeguard the health of living beings [31].

Biocides and antifungal agents such as Tributyltin (TBT) are commonly employed in various consumer products like footwear, socks, and athletic apparel to prevent unpleasant odors resulting from sweat decomposition. Tributyltin is classified as a *priority hazardous substance* under European Union (EU) guidelines, necessitating the adoption of measures to eradicate its contamination of surface waters in Europe. The EU has prohibited products, including consumer goods, containing over 0.1% of specific organotin compounds [32]. Additionally, the application of chemicals such as DEET (N, N-diethyl-m-toluamide), DDT (Dichlorodiphenyltrichloroethane), permethrin, and others in the textile finishing process for insect/mosquito repellency has been associated with adverse effects on the users' physiological well-being and overall health48. Consequently, there is a growing recognition within the scientific community for the adoption of sustainable chemicals and naturally derived substances, particularly those sourced from plants or herbal extracts, in the development of insect repellent textile finishes [46, 47].

Soil and stain release treatments are especially beneficial in situations where the market calls for white or light-colored clothing or when engaging in leisure or sports activities that may result in higher levels of dirt or stains. These treatments must be able to withstand washing, and there are three primary types available: (1) treatments with carboxyl groups; (2) treatments with oxyethylene and/or hydroxyl groups; and (3) fluorocarbon treatments [30].

Fabrics and clothing that have a pleasant texture are highly valued for their comfortable feel. However, they can also affect the way the fabric drapes, is cut, and sewn. Chemical softeners work by lubricating the fibers and reducing friction between them and metal surfaces. Today's consumers demand softeners that are fast-acting, prevent static buildup on hydrophobic fibers, and maintain the textile's ability to absorb moisture. Depending on the type of softener used, they may also include antistatic agents, bactericidal properties, and soil release properties. The classification of fabric softeners primarily encompasses three categories: anionic, cationic, and nonionic compounds. It is worth noting that certain fabric softener products may incorporate a combination of nonionic and cationic compounds, while others may possess reactive softening agents. Softening formulations often incorporate a diverse range of chemicals to achieve their desired effects. These chemicals encompass tetrahydropyrimidinium, and quaternary ammonium salts containing reactive ester (RCOOR′) groups, amine oxides, and quaternary amine-boric acid mixtures, among other compounds [30, 33].

The process of dry cleaning involves using a chemical solvent instead of water to clean clothing and textiles. This method is particularly useful for delicate fabrics or those with low wet-fastness dyes. Tetrachloroethylene is the most extensively employed solvent; it is often referred to as PERC, or perchloroethylene. However, it has been categorized as a carcinogen by the United States Environment Protection Agency, and any waste products must be treated as hazardous material to avoid contaminating potable water. Hence, dry cleaners must exercise heightened caution to mitigate any potential adverse impacts on the environment [34].

Octylphenols (OPs) and nonylphenols (NPs), along with their corresponding ethoxylates, notably Nonylphenol Ethoxylates (NPEOs), have found extensive use in the apparel sector, primarily in laundering and dyeing procedure. However, these compounds are known to pose significant harm to marine life. Alkylphenol Ethoxylates (APEOs), including Nonylphenol Ethoxylates (NPEOs), are recognized as endocrine disruptors and have been linked to the feminization of male fish. Since 2005, the European Union has placed strict limitations on the distribution of goods that contain over 0.1% of NPEOs. Additionally, Chlorophenol compounds are employed as biocides within the textile sector, with Pentachlorophenol (PCP) being particularly toxic to marine animals and potentially harmful to human vital organs and the central nervous system. The European Union (EU) has implemented a ban on the production and utilization of Phencyclidine (PCP) since 1991 [35].

4 Utilizations of Enzymes in the Textile Sector

In the past, textile manufacturing stages like desizing, scouring, and bleaching relied on chemicals such as urea, sodium hydroxide, salts, acids, bases, and so on. These substances, while effective in their respective roles, possess inherent toxicity to human health and can contribute to environmental pollution upon disposal. Consequently, enzymes have appeared as a cleaner and more sustainable substitute. Enzymes find two primary applications in the apparel industry: amylases are used in desizing process, and cellulases are employed for the bio-stoning and softening of cotton goods. Additionally, textile processing utilizes enzymes like catalases, lipases, proteases, pectinases, xylanases, and others for various processes, including bio-polishing, denim fading, bio-scouring, removal of peroxide, wool finishing, and more.

Enzymes have become the preferred choice over traditional chemicals in textile industries due to several advantages. They speed up reaction rates, exhibit high degrees of substrate specificity, and operate under milder environments, thereby affording greater ease of control. Furthermore, enzymes possess the advantageous attributes of biodegradability and enhanced safety, rendering them viable alternatives to the utilization of hazardous chemicals [36, 37]. Table 3 provides a list of various enzymes commonly used in the textile industry.

Table 3 List of major enzymes employed in the fashion industry [36, 38, 39]

Enzymes	Process
Amylases	Desizing of cotton
Sericinases	Degumming of silk
Catalase	Clean remnant bio-bleaching solution
Cellulase	Bio-stoning of jeans, bio-polishing
Collagenase	Surface modification of wool
Cutinase	Bio-scouring
Esterase	Bio-bleaching and modification of surface properties
Pectinase	Bio-scouring cotton and jute
Laccase	Bio-bleaching, textile effluent decolorization
Peroxidase	Bio-bleaching
Protease	Bio-scouring, silk degumming, surface modification of wool
Transglutaminase	Shrinkage reduction in woollen fabric

5 Eco-friendly Alternatives to Hazardous Chemicals in Textile Wet-Processing

In light of the negative consequences associated with the application of synthetic chemicals in the wet processing of textiles, a variety of environmentally friendly alternatives have been introduced as a means of mitigating these adverse effects. Table 4 presents an extensive list of environmentally sustainable alternatives to traditional chemicals and treatments within the textile industry. In addition, it is worth noting that a multitude of international organizations are diligently striving to mitigate the risks and deleterious consequences stemming from the utilization of chemicals in the textile sector. Table 5 presents a comprehensive compilation of legislations that have been formulated by various organizations specifically targeting consumer goods. The primary objective of these endeavors is to advance the adoption of practices that enhance safety and environmental sustainability within the textile industry.

Table 4 List of nature-friendly substitutes or processes for wet-processing textiles [4, 40–45]

Stages	Textile wet processes	Chemicals used	Alternatives
Preparatory process	Sizing	Starch	Water-soluble polyvinyl alcohol
	Desizing	Hydrochloric acid	Amylases (enzyme)
	Desizing PVA	–	Ultraviolet C
	Scouring	Sodium hydroxide	Pectinases (enzyme)
	Neutralization	Acetic acid	Formic acid
	Bleaching	Hypochlorite	Hydrogen peroxide
	Bleaching	Hydrogen peroxide	Laccases (enzyme)
	Bleaching	Hydrogen peroxide	Peracetic acid (PAA)
	Peroxide killer	Sodium thiosulfate	Catalases (enzyme)
	Mercerization	Sodium hydroxide	Liquid ammonia
Textile coloration (dyeing and printing)	Conventional water dyeing: disperse dyes	Water and other auxiliaries	Supercritical CO_2 dyeing (waterless dyeing)
	Dyeing (silk and wool)	Acid dyes	Microbial dyes
	Oxidation of VAT and sulfur dyes	Hydrogen peroxide, sodium perborate	Potassium dichromate
	Sulfur dyeing	Powder form of sulfur dyes	Pre-reduced dyes
	Thickener	Kerosene	Water-based polyacrylate copolymers
	Printing	Conventional printing	Digital printing (drop-on-demand technology)
Finishing	Water repellent	C8 fluorocarbons	C6 fluorocarbons
	Crease recovery	Formaldehyde-based resin	Polycarboxylic acid
	Wetting agents, detergent	Alkyl phenol ethoxylates	Fatty alcohol phenol ethoxylates
	Shrink proofing	Chlorination	Plasma treatment
	Flame retardant	Bromated diphenyl ethers	Combination of phosphonates and inorganic salts
	Hydrotropic agent	Urea	Dicyanamide (partially)

Table 5 Laws and regulations governing chemicals in consumer products [30]

Legislation	Country
Biocidal Product Regulation	European Union
Canadian Environmental Protection Act	Canada
Chemical Substances Control Law	Japan
Children's Safe Product Act	USA
Consumer Product Safety Improvement Act	USA
Registration Evaluation Authorization and Restriction of Chemicals	European Union
Regulations on Limitations of Substances in Products	Norway
Self-Regulatory Safety Confirmation Act	Korea ara>
The Chemical Risk Reduction Ordinance	Switzerland
The Safe Drinking Water and Toxic Enforcement Act of 1986 (California Proposition 65)	USA
Toxic Substances Control Act	USA

6 Conclusion

While significant progress has been made in eliminating harmful substances from textile processes, the presence of textile chemicals known as wet-processing aides remains a challenge. These products are proprietary blends composed of intricate combinations of softeners, chelating agents, solvents, surfactants, as well as water-based polymers. They are specifically engineered to perform crucial functions in textile preparation, coloring, and finishing. The complexity arises from the vast array of chemicals that can be used and the undisclosed concentrations employed in these blends, which manufacturers often guard as trade secrets. The pivotal question revolves around how to assess the environmental impact of these products. This knowledge is vital for textile industry consumers to make informed choices, opt for eco-friendly alternatives, and contribute to improving the water quality of effluents discharged from textile plants. Despite these challenges, several prominent global corporations, including Adidas, H&M, and Nike, are actively engaged in efforts to reduce the impact of harmful substances on water bodies. They are committed to implementing eco-friendly and sustainable chemical alternatives, demonstrating a positive shift towards a more environmentally responsible textile industry.

References

1. Park JH, Glouberman E (1932) The importance of chemical developments in the textile industries during the industrial revolution. J Chem Educ 9(7):1142. https://doi.org/10.1021/ed009p1142
2. Kan CW (2015) Plasma treatments for sustainable textile processing. In: Sustainable apparel. Woodhead Publishing, pp 49–118. https://doi.org/10.1016/B978-1-78242-339-3.00003-0

3. Amutha K (2017) Sustainable chemical management and zero discharges. In: Sustainable fibres and textiles. Woodhead Publishing, pp 347–366. https://doi.org/10.1016/B978-0-08-102041-8.00012-3
4. Gulzar T, Farooq T, Kiran S, Ahmad I, Hameed A (2019) Green chemistry in the wet processing of textiles. In: The impact and prospects of green chemistry for textile technology. Woodhead Publishing, pp 1–20. https://doi.org/10.1016/B978-0-08-102491-1.00001-0
5. Roy Choudhury AK (2014) Environmental impacts of the textile industry and its assessment through life cycle assessment. Roadmap to sustainable textiles and clothing: environmental and social aspects of textiles and clothing supply chain, pp 1–39. https://doi.org/10.1007/978-981-287-110-7_1
6. Sule AD, Bardhan MK (1999) Objective evaluation of feel and handle, appearance and tailorability of fabrics Part-II: The KES-FB system of Kawabata. Colourage 46(12):23–28
7. Khan S, Malik A (2014) Environmental and health effects of textile industry wastewater. Environmental deterioration and human health: natural and anthropogenic determinants, pp 55–71. https://doi.org/10.1007/978-94-007-7890-0_4
8. Azanaw A, Birlie B, Teshome B, Jemberie M (2022) Textile effluent treatment methods and eco-friendly resolution of textile wastewater. Case Stud Chem Environ Eng:100230. https://doi.org/10.1016/j.cscee.2022.100230
9. American Apparel & Footwear Association. (n.d.). Product safety guidance: restricted substances list (RSL). https://www.aafaglobal.org/AAFA/Solutions_Pages/Restricted_Substance_List
10. Choudhury AKR (2017) Sustainable chemical technologies for textile production. In: Sustainable fibres and textiles. Woodhead Publishing, pp 267–322. https://doi.org/10.1016/B978-0-08-102041-8.00010-X
11. Anonymous (2020, September 23) EU countries agree textile chemical ban. The Guardian. https://www.theguardian.com/environment/2015/jul/21/eu-countries-agree-textile-chemical-ban
12. Hasan M, Nabi F, Mahmud R (2015) Benefits of enzymatic process in textile wet processing. Int J Fiber Text Res 5(2):16–19
13. Ministry of Chemicals and Fertilizers (2021) Annual report 2020–21. Department of Chemicals and Petro-chemicals, Ministry of Chemicals and Fertilizers, Government of India. https://chemicals.gov.in/sites/default/files/Reports/Annual_Report_2020-21.pdf
14. Ministry of Chemicals and Fertilizers (2022) Annual report 2021–22. Department of Chemicals and Petro-chemicals, Ministry of Chemicals and Fertilizers, Government of India. https://chemicals.gov.in/sites/default/files/Reports/Annual_Report_2021-22.pdf
15. Goswami BC, Anandjiwala RD, Hall D (2004) Textile sizing, 1st ed. CRC Press. https://doi.org/10.1201/9780203913543
16. Kiron MI (2012) Desizing process in textile: objectives and methods. Textile Learner. https://textilelearner.net/desizing-process-in-textile-objectives-and-methods/
17. Purushothama B (2019) Handbook of value addition processes for fabrics. Woodhead Publishing India Pvt Limited
18. Shenai VA (1987) Technology of textile processing. Vol-II, Chemistry of Dyes. Sevak Publications
19. Athalye A (2014) Bleach clean up. FIBRE2FASHION. Retrieved April 17, 2023, from https://www.fibre2fashion.com/industry-article/7349/bleach-clean-up
20. John MJ, Anandjiwala RD (2009). Surface modification and preparation techniques for textile materials. In: Surface modification of textiles. Woodhead Publishing, pp 1–25. https://doi.org/10.1533/9781845696689.1
21. Choudhury AKR (2011). Pre-treatment and preparation of textile materials prior to dyeing. In: Handbook of textile and industrial dyeing. Woodhead Publishing, pp 64–149. https://doi.org/10.1533/9780857093974.1.64
22. Shang SM (2013) Process control in dyeing of textiles. In: Process control in textile manufacturing. Woodhead Publishing, pp 300–338. https://doi.org/10.1533/9780857095633.3.300

23. Choudhury AKR (2022) Principles of textile printing. CRC Press, Boca Raton, pp 374. https://doi.org/10.1201/9781351067836
24. Tyler DJ (2005) Textile digital printing technologies. Text Prog 37(4):1–65. https://doi.org/10.1533/tepr.2005.0004
25. Goel PK (2006) Water pollution, causes, effects and control. New Age International (P) Ltd. Publishers, New Delhi. https://books.google.co.in/books?id=4R9CYYoiFCcC&source=gbs_similarbooks
26. Outdoor Industry Association (2012) Durable water and soil repellent chemistry in the textile industry—a research report. In: P05 Water repellency project. Retrieved April 18, 2023, from https://outdoorindustry.org/wp-content/uploads/2015/05/FINAL_ZDHC_P05_DWR-Research_Nov20121.pdf
27. Holmquist H, Schellenberger S, van Der Veen I, Peters GM, Leonards PEG, Cousins IT (2016) Properties, performance and associated hazards of state-of-the-art durable water repellent (DWR) chemistry for textile finishing. Environ Int 91:251–264. https://doi.org/10.1016/j.envint.2016.02.035
28. Organisation for Economic Co-operation and Development (2002) Hazard assessment of perfluorooctane sulfonate (PFOS) and its salts. Retrieved April 18, 2023, from https://www.oecd.org/chemicalsafety/risk-assessment/2382880.pdf
29. Nimkar U (2018) Sustainable chemistry: a solution to the textile industry in a developing world. Curr Opin Green Sustain Chem 9:13–17
30. Holme I (1993) New developments in the chemical finishing of textiles. J Text Inst 84(4):520–533. https://doi.org/10.1080/00405009308658985
31. Ericsson (2012) The Ericsson lists of banned and restricted substances, 2/000 21-FAU 104 04 Uen Rev C, 15 June 2012. http://www.ericsson.com
32. The Commission of the European Communities (2009) Amending council directive 76/769/EEC as regards restrictions on the marketing and use of organostannic compounds for the purpose of adapting its Annex I to technical progress. Official Journal of the European Union. https://eur-lex.europa.eu/LexUriServ/LexUriServ.do?uri=OJ:L:2009:138:0011:0013:EN:PDF
33. Roy Choudhury AK (2017) Softening. In: Principles of textile finishing. Woodhead Publishing Ltd., London, UK, pp 109–148. https://doi.org/10.1016/B978-0-08-100646-7.00006-0
34. United States Environment Protection Agency (2013) Chemicals in the environment: perchloroethylene. US Environmental Protection Agency, Office of Pollution Prevention and Toxics 1994–2008. https://www.epa.gov/assessing-and-managing-chemicals-under-tsca
35. Greenpeace International (2011) Dirty laundry. https://www.greenpeace.org/international/publication/7168/dirty-laundry/
36. Mojsov K (2011) Application of enzymes in the textile industry: a review. In: Conference: II international congress: engineering, ecology and materials in the processing industry, pp 230–239
37. Madhu A, Chakraborty JN (2017) Developments in application of enzymes for textile processing. J Clean Prod 145:114–133. https://doi.org/10.1016/j.jclepro.2017.01.013
38. Al-Sayed W, Abdelrahman SH (2021) Sustainable chemistry in textile processes (pretreatment, coloration and chemical finishing). In: Green chemistry for sustainable textiles. Woodhead Publishing, pp 93–111. https://doi.org/10.1016/B978-0-323-85204-3.00022-1
39. Sen A, Kapila R, Chaudhary S, Nigam A (2021) Biotechnological applications of microbial enzymes to replace chemicals in the textile industry-a review. Text Asso 82(2):68–73
40. Unuofin JO, Moloantoa KM, Khetsha ZP (2022) The biobleaching potential of laccase produced from mandarin peelings: impetus for a circular bio-based economy in textile biofinishing. Arab J Chem 15(12):104305. https://doi.org/10.1016/j.arabjc.2022.104305
41. Khan F (2010) A comparison of peracetic acid and hydrogen peroxide bleaching on cotton fabric. Fibre2fashion. https://www.fibre2fashion.com/industry-article/5055/a-comparison-of-peracetic-acid-and-hydrogen-peroxide-bleaching-on-cotton-fabric
42. Križman P, Kovač F, Tavčer PF (2005) Bleaching of cotton fabric with peracetic acid in the presence of different activators. Color Technol 121(6):304–309. https://doi.org/10.1111/j.1478-4408.2005.tb00373.x

43. Ramsey E, Qiubai S, Zhang Z, Zhang C, Wei G (2009) Mini-review: green sustainable processes using supercritical fluid carbon dioxide. J Environ Sci 21(6):720–726. https://doi.org/10.1016/S1001-0742(08)62330-X
44. Banchero M (2013) Supercritical fluid dyeing of synthetic and natural textiles–a review. Color Technol 129(1):2–17. https://doi.org/10.1111/cote.12005
45. Verma H, Sharma AK, Rani A, Thiyam G, Pavan M (2023) Optimization of the process parameters for dyeing silk fabric with a fungal pigment using RSM. Sustain Chem Pharm 35:101192. https://doi.org/10.1016/j.scp.2023.101192
46. Parvez AAl, Hossain MJ, Hossain MZ, Sohan MSH, Hoque F, Ahsan MH, Hoque MS (2023) Mosquito repellent fabric: Development and characterization of peppermint and garlic mixture finish on knitted fabric to examine mosquito repellency. Heliyon 9(5):e15944. https://doi.org/10.1016/j.heliyon.2023.e15944
47. Pavan M, Rani A, Jhang T, Singh SP (2023) Developing herbal mosquito repellent cotton fabric using the optimized process variables for the safe environment. Mat Today: Proc 78:900–906. https://doi.org/10.1016/j.matpr.2022.12.191
48. Sajib M, Banna B, Mia R, Ahmed B, Chak R, Alam SS,... Islam MT (2020) Mosquito repellent finishes on textile fabrics (woven & knit) by using different medicinal natural plants. J Text Eng Fashion Tech 6(4):164–167. https://doi.org/10.15406/jteft.2020.06.00246

Water Consumption and Microfibers: The Biggest Threat

Oinam Roselyn Devi and Laimayum Jogeeta Devi

Abstract The textile industry is one of the top industries that use enormous amounts of water. Water is used extensively throughout textile wet processing operations contributing to the release of industrial wastewater. It is becoming critically essential for textile operations to secure a cost-effective and reliable water supply. To lessen the water footprint, the use of supercritical fluids and an ultrasound-assisted method has also been explored in the textile manufacturing process. The potential of plasma technology, laser finishing, UV radiation treatment, gamma (Y) ray irradiation, and ozone application to enhance functional and aesthetic finishing, such as dyeing and hydrophilic and hydrophobic finishing, has been the focus of numerous research efforts recently. Microfiber pollution is also another big threat to biodiversity from the textile sector. Washing clothes annually releases 50 billion pounds of plastic into the ocean or around 500,000 tonnes of microfibers. Due to fast fashion and the rise in population, textile production and consumption of cheap fabrics made of synthetic fibers have increased. It has been estimated that up to 35% of the primary source microplastics in the marine environment come from synthetic clothing. Natural fibers also contribute to microfiber pollution, but it is biodegradable and has minimal environmental impact. However, traces of chemicals and dyes may present in natural and semi-synthetic microfibers which are equally problematic contaminants. In recent studies, microfibers have been reported in freshwater and marine environments and even in several products for human consumption including tap water, beer common salt, and seafood. Hence, microfiber pollution becomes a serious threat to the environment and it needs to be addressed. Research on the scope of impacts, ecological implications, and potential health effects of microfibers on humans is required. The chapter also provides insight into microfiber pollution, threats, and management strategies.

O. R. Devi (✉)
Department of Apparel and Textile Science, Punjab Agricultural University, Ludhiana 141004, India
e-mail: oinam.roselyn@gmail.com

L. J. Devi
Department of Textile and Apparel Designing, Pandit Deen Dayal Upadhyay Institute of Agricultural Sciences, Imphal, Manipur, India

Sadhna et al. (eds.), *Climate Action Through Eco-Friendly Textiles*, SDGs and Textiles,
https://doi.org/10.1007/978-981-99-9856-2_6

Keywords Microfiber · Microplastic · Pollution · Threat · Water consumption

1 Introduction

The textile sector ranks among the top ten in terms of water consumption. The entire textile wet processing process, which entails scouring, bleaching, dyeing, printing, and finishing, requires a significant amount of water. The precise methods, equipment, and approaches to managing water use in the textile sector establish the overall volume of water utilized. A process house uses nearly three times as much water as the combined usage of all other units. In terms of the various natural textiles, the processing of wool and other felted fabrics required more water than those for making woven, knit, stock, and carpet. About 95% of the water required in cotton textile production goes towards producing the raw materials, with the remaining 5% going into the processing of cotton fabric. For the preparation of 1 kg of cotton textile, 100–150 L of water is needed. Usually producing a pair of cotton jeans uses more water (greater water footprint). Wool and silk, protein fibers also require more water due to scouring and coloring. According to recent reports, bleaching cellulosic fabric uses more water than other techniques. In terms of the detailed dyeing process, different classes of colors require varying amounts of water. Additionally, the quantum of water utilized varies as per the dyeing machine and procedures used. Compared with dyeing, coloring via printing technology significantly requires less water. In this regard, the development of digital printing is noteworthy because it is a completely water-free technique. According to industry studies, jigger dyeing equipment uses around three times as much water as beam and beck dyeing equipment; that difference could be attributed to the higher liquid-to-goods ratio.

Recently, several textile dyers have started adopting pad-dry-pad-steam (pretreatment range) and continuous dyeing machines to reduce the amount of water required for industrial production. To lessen the water footprint, the use of supercritical fluids and an ultrasound-assisted method has also been explored in the textile manufacturing process. Given the scarcity of clean water, the world has been developing methods for conserving water in the past few decades. Another major problem that needs to be appropriately addressed is the effluent produced by textile industries.

In the past, a suitable treatment plant facility was considered to be established for discharging contaminated water effluents without damaging agriculture and aquaculture, or to totally or partially utilize the processed water. The levels of several environmentally hazardous dyes, pigments, finishing chemicals, auxiliary substances, dispersion agents, and other substances have raised the levels of pollution indicator substances such as total dissolved solids (TDS), chemical oxygen demand (COD), and biochemical oxygen demand (BOD) in water. Many researchers have worked to prove that aesthetic and functional finishing techniques like dyeing and hydrophilic and hydrophobic finishing can be improved by plasma technology, laser finishing, UV radiation treatment, gamma (Y) ray irradiation, and ozone application [1]. Numerous water and solvent-based textile coatings with thicknesses ranging from the micron to sub-micron size have also been explored together with several of these advantageous modifications to textiles, leading to a decrease in water consumption in the textile industry. Additionally, it aids in the development of functional textiles with the appropriate functionality by altering the material's entire surface at the micron to the nanoscale. These irradiation methods have the potential to significantly quicken the dyeing process and increase dye uptake in all fibers, synthetic and natural, in less time.

2 Water Usage in Textile Industry

The textile sector is one of the largest consumers of water in the manufacturing process and a significant source of industrial effluent. Since different textile operations involve a variety of chemicals, the wastewater from those processes contains many harmful substances that, if not adequately handled before being generated into the environment, can seriously harm the ecosystem [2].

2.1 Water Usage During Natural Fiber Processing

In the textile industry, water is widely used to process cellulosic fibers like cotton, flax, bamboo, nettle, and ramie, as well as proteinous fibers like silk and wool. Water is used extensively in the cotton textile industry to desize, scour, bleach, dye, print, and finish size warp yarn. Additionally, water is needed for washing and rinsing the processed materials. A rough estimate of the water used in the manufacturing of cotton ranges from 250 to 350 L/kg of fabric and 200 to 300 L/kg of fabric for wool. Table 1 displays the amount of water used in the manufacturing of different types of fibers. Generally speaking, a process house uses three times as much water as any other input.

Table 2 lists the water required for textile wet processing, from desizing to bleaching, together with the requirement of water in terms of washing phases. It has been noted that a significant amount of water is needed for the production of textiles. However, it differs from one mill to another, as per the kind of textile materials used, the equipment, the process, and the dyes used. The washing machine uses a lot of water after each procedure, as illustrated in Table 2. Additionally, a lot of water is needed for the production of natural yarns such as cotton, linen, and hemp. Water usage in the various production phases varies significantly due

Table 1 Water utilization for the manufacturing of various textile fibers [3]

Various textile fibers	Cotton	Wool	Nylon	Rayon	Polyester
Water utilized (L/Kg of fabric)	250–350	200–300	125–150	125–150	100–200

Table 2 Water consumption and energy in Kiers and J-Box processes [3]

Process	Desizing	Washing	Scouring	Washing	Bleaching	Washing	Total
Consumption of water L/kg (Kier's process)	3	20	2	20	2	40	87
Consumption of steam kg/kg (J-Box Processes)	0.25	0.35	1.75	0.30	1.00	0.60	4.20

Table 3 Liquor to good ratios required in various dyeing equipment [4–7]

Dyeing machine	Continuous	Winch	Jet	Jigger	Beam	Package	Beck	Stock	Skein (Hank)
Liquor to good ratios	1:1	15:1–40:1	7:1–15:1	5:1	10:1	10:1	17:1	12:1	17:1

to the use of newer equipment, such as a continuous dyeing machine and the pad-dry-pad-steam (pretreatment process). The material-to-liquor ratio (MLR ratio) for processing natural fibers varies depending on the dyeing equipment used in wet processing. Water, chemicals, and energy might all be significantly reduced by a low-liquor-ratio dyeing process. Table 3 displays the liquor ratios for several types of dyeing equipment together with the normal water requirements for dyeing.

Data from Table 3 indicate that continuous dyeing equipment uses less water than other dyeing machines. Many textile firms employ modern continuous textile processing machinery, which uses very little water. Due to their high initial costs, limited processing time, and incompatibility with knitwear and crepe textiles, such water-efficient machines can be challenging to install [5].

2.2 Water Usage During Processing of Synthetic Fibers

The total water consumption of textiles made of synthetic fibers is shown in Table 4. Water is used less when processing synthetic textiles than when processing natural textiles. Polyester can be processed wet with just 100–200 kg/kg of water. The type of machinery, the kind of fiber, and the application of both contemporary and antiquated technology all have a big influence on water consumption. Table 4 also shows how much water is required at each stage.

Table 4 Water usage for processing of synthetic fibers [3]

Wet Processing	Water consumption in L/1000 kg fibers				
	Rayon	Acetate	Nylon	Acrylic/ Modacrylic	Polyester
Scouring	1700–34,000	25,000–84,000	50,000–67,000	50,000–67,000	25,000–42,000
Salt bath	4000–12,000	–	–	–	–
Bleaching	–	33,000–50,000	–	–	–
Dyeing	1700–34,000	34,000–50,000	17,000–34,000	17,000–34,000	17,000–34,000
Finishing	4000–12,000	24,000–40,000	32,000–48,000	40,000–56,000	8000–12,000

3 Strategies for Reducing Water Consumption in Wet Processing

One of the biggest problems the textile industry has is the creation of textile effluents, which must be properly managed. The use of plasma pretreatment, supercritical dyeing, biological bleaching (Biobleaching), foam finishing, single-stage preparatory processes, ultrasonic and microwave dyeing, and electrochemical dyeing are some of the technical developments made in this regard, i.e., to reduce the amount of wastewater generated during textile wet processing [8–13].

3.1 Use of Ultrasonic Wave-Assisted in Textile Processing

The textile wet process sequence, desizing, scouring, bleaching, mercerization, dyeing, and finishing use the ultrasonic wave-assisted approach [14–16]. The use of ultrasonic techniques for desizing is more cost-effective than traditional methods. Additionally, it has been found that ultrasonic treatment completely removes oil from cotton and nylon materials. This technique has no impact on the stability or the characteristics of the fiber. After bleaching, ultrasonic (20 kHz) treatment considerably increases the bleaching (peroxide) rate and the fabric's whiteness index. Additionally, by lowering the processing temperature, the technique aids in energy conservation. When compared to conventionally treated fabric, ultrasound-treated fabric exhibits greater fastness and absorbency. When mercerizing 100% cotton fabric, this technique accelerates the process by two to three times.

Al-Etaibi and El-Apasery dyed the polyester fabric with dispersed dye using ultrasonic energy [17]. It has been demonstrated that the ultrasound dyeing technique has better color fastness than the traditional dyeing method because ultrasonic waves speed up the diffusion of dyes into materials and also increase the dyeing rate. Polyester dyed with dispersion dyes employing an ultrasonic approach has a higher UV protection factor (UPF) than polyester dyed using traditional dyeing techniques.

3.2 Different Irradiation Techniques for the Processing of Textiles

The introduction of ultraviolet (UV) and plasma irradiation technology has made it possible to process textiles in a way that uses less water, energy, and chemicals. It has been noted that UV-treated wool fabrics improved dyeing rates as well as saturated color uptake. This happens because the UV treatment in the nitrogen environment increases the number of amine groups on the wool surface at the nanoscale, which increases the number of dye sites available for anionic (acid) dye molecules. Additionally, it has been noted that UV radiation can cause free radicals to be generated, which modify the surface scales of wool, and increase the dye uptake. These effects can all be seen on the wool surface. Additionally, UV irradiation shortens the time and temperature of the dyeing process, which speeds up the wool dyeing process and uses less energy than traditional dyeing methods [18, 19].

As an alternative to dangerous chlorinated chemicals, UV and plasma technologies have also been investigated to remove scales from the wool surface (antishrink finish). Plasma technology, like UV pretreatment, aids in shortening the dyeing cycle and speeding up color absorption. Additionally, either in situ or post-polymerizing hydrocarbon or fluorocarbon precursors, cotton fabrics have been rendered hydrophobic using plasma [20]. However, the consistency of the dyeing and other functional finishes is a key drawback of this irradiation technique. Furthermore, because most of the dye molecules adhere to the outer layer of the substrate, colored samples dry sufficiently. While gamma and laser radiation have also been demonstrated to be effective in textile processing, their high initial costs and uniformity of treatment hinder their general application and success.

3.3 Waterless/Low Water Processing Using Supercritical Fluid

Most substances are found in one of three states: solid, liquid, or gas. Molecules are dispersed at random throughout the system during the gaseous phase. Certain fluids (like gas) change into supercritical liquid, which is neither a gas nor a liquid, when pressure and temperature rise above a critical point. Water pollution is successfully avoided during the production process by using this type of liquid. Carbon dioxide is especially popular in this situation because it is inexpensive and readily available in the environment. It is not only inexpensive but also non-poisonous, chemically unreactive, recyclable, renewable, and environmentally benign with zero waste formation. Its lower surface tension, lower viscosity, and handiness are some other advantages [8–13]. The process of dyeing fabric with a supercritical fluid involves encasing the cloth in a perforated stainless steel tube and subjecting it to autoclaving. The dye is positioned at the chamber's bottom, and the temperature is raised by constantly exhaling carbon dioxide while starting continuously. Pressure is raised after the temperature reaches the critical point, and both are then maintained for 60 min during the dyeing cycle. The residual dyestuff and carbon dioxide are recycled when the operation is finished under pressure. The dyed dried sample is rinsed with acetone to get rid of any remaining dyestuff on the fabric surface. Compared to traditionally colored fabric, the dyeing process produces a high degree of levelness since it occurs in a virtual gas phase.

3.4 *Nanocoating of Textiles*

Various kinds of synthetic polymers such as polyvinyl chloride (PVC), nylon, thermoplastic co-polyester, polyolefin, polyurethane, polyacrylate, polyvinyl alcohol (PVA), and polyvinyl acetate are used to give the fabric functional properties [8–14]. A substrate may sometimes also be coated with pressure-sensitive adhesive. Natural rubber, acrylic polymer, and styrene butadiene rubber are the three most commonly used adhesives. A solvent-based adhesive can be used to cover the textile material with a microporous membrane that repels water. This type of coated cloth is commonly found in hospital operating rooms. Another method for reducing water use is powder coating. The fabric is coated with polymer granules ranging in size from 60 to 200 m, which are then heated to melt the granules. It is then covered with another layer of cloth before the liquid polymer solidifies. No water or solvents are used in this process. A textile substrate can be coated with a nanocoating using layer-by-layer (LbL) deposition, chemical vapor deposition (CVD), or plasma spraying. This type of coating does not affect the appearance or feel of the cloth. Even though this method has several benefits, during mass production, non-uniformity appears all over the surface.

3.5 *Spray and Foam Finishing of Textiles*

From the perspective of water conservation, the spray technique in textile finishing was developed. For drying and curing, this process also consumes less energy, water, and chemicals. This process can be used in several functional finishes (antibacterial, softness, hardness, and water repellent) [8–13].

Another method for reducing water usage and processing costs is foam finishing. It can be made by continuously forcing air into a liquid that has different chemicals and foaming agents in it. Air bubbles scattered in a liquid medium make up the colloidal system, known as foam. The density of the foam bubbles and the amount of non-foamed liquor used affect the viscosity of the foam. Foam can be made more stable by increasing viscosity and decreasing droplet size.

3.6 Printing for Textile Coloration

The traditional printing method is time-consuming and labor-intensive. Additionally, the procedure requires a lot of water and power to run. To reduce some of the drawbacks of conventional printing methods, like water and electricity use, digital printing has entered the market to address these issues. The most common color used is CMYK (cyan-magenta-yellow-black). A continuous ink jet printing system offers an extremely fast rate of output and no color restrictions. Another most recent printing method is drop-on-demand technology. This technology has the most sophisticated system, and much study is being conducted in this area to improve it further. This technology is further divided into categories based on the ink jet's electrostatic and acoustic characteristics, temperature behavior, and valve type. A dye or pigment could be used as printing ink. To ensure bleeding resistance, the fabric is first chemically treated. In 1975, Milliken introduced the Millitron digital printing system, which prints carpet and upholstery using the continuous valve method. The first ink jet printing device used for commerce was the Stork "True color" model. Ciba Specialty Chemicals, Scitex Vision Ltd., and Reggiani developed the DReAM device. Industrial manufacturing on a wide scale is perfect for this machinery. The heating and cleaning processes were dropped in later advancements in this area. Recently, the pad, dry, and cure method has been used to demonstrate pigment-based ink jet printing on cotton fabric [22–24].

4 Microfiber

Another threat to biodiversity from textile sectors is microfiber pollution. Worldwide, freshwater and marine environments have been shown to contain novel pollutants called microfibers. The ocean contains about 1.5 million trillion microfibers, and each year, 2 million tonnes of new microfibers are produced and discarded into the water [25]. According to estimates, up to 35% of all primary source microplastics in the marine environment are caused by synthetic clothing [26]. According to Carr, more than 85% of the microplastic debris discovered along the world's shorelines is composed of microfibers [27]. The consensus was that household laundry was the primary source of microfiber pollution, even though the fundamental causes have not been thoroughly examined. Viscose made up 56.9% of all microfibers in deep-sea sediments from the Atlantic Ocean floor, which is more than twice as much as polyester [28]. 34.8% of the generation of major microplastics into the water can be attributed to washing synthetic garments [29]. Polyester was found to be the most common type of microfiber in both saltwater and marine sediments, with concentrations ranging from 13.15 to 39.48 items per 250 g dry weight and 503 to 459,681 items km^{-2}, respectively [30].

4.1 Microfiber Pollutant

Microfiber is a man-made fiber finer than one denier that has a diameter of less than 10 m. Another notable feature of microfibers is their high length-to-diameter ratio, which is roughly 103. In textile engineering, a microfiber is sometimes identified as a man-made fiber; however natural fibers are also included [31]. Research suggests that natural fibers predominate over synthetic fibers in freshwater and airborne microfibers, but due to their biodegradability, natural fibers are expected to have minimal effects on ecosystems and humans [32]. Indeed, further research is needed to assess the degree of the effects, ecological implications, and health risks associated with microfibers in humans [33].

4.2 Source of Microfiber

Due to fast fashion and population increase, textile production and consumption are always rising. Synthetic fibers that are one denier or finer are called microfibers. A microfiber's diameter is less than that of a silk strand and is around one-fifth that of human hair. According to Bhatt and Rani [34], the benefits or features of these microfibers are their softness, absorbency, water repellency, electrostatics, and filtering abilities. When synthetic clothing is laundered, microscopic microfibers of plastic fall off. Microfiber is the most prevalent kind of microplastic; it has a diameter of less than 5 mm and is undetectable to the human eye. They have been discovered in the environment as a result of the degradation of larger plastic objects such as bottles, bags, exfoliating scrubs, toothpaste, and fishing equipment [35]. It is estimated that washing clothes annually releases 50 billion pounds of plastic into the ocean, or around 500,000 tonnes of microfibers [34]. According to research, one piece of clothing discharges more than 1900 microfibers in just one time wash [36]. Recent discoveries have also revealed that natural and semi-synthetic microfibers are equally dangerous pollutants because of their ability to release a wide range of chemicals and dyes.

Numerous studies indicate that one of the main causes of microfiber pollution is washing garments. According to an Ocean Wise study report from 2019, the average household in the United States and Canada is estimated to release 135 g or 533 million microfibers annually into wastewater treatment plants. Once they get into the environment, microfibers are very difficult to remove [37]. India, one of the world's leading plastic producers, experienced a significant increase in the contamination of microplastics. Therefore, it is essential to stop this microfiber pollution from contaminating the ecosystem [38].

4.3 Classification of Sources of Microplastics

Major sources for the release of microplastics can be classified into the following:

1. Personal care items with microbeads
2. Synthetic clothing
3. Packaging materials
4. Laundry
5. Non-clothing textiles
6. Items of degraded marine litter.

4.4 Effects/Threats of Microfiber Pollution

Microfiber pollution may have potentially severe consequences and negative impacts on the environment and humans. Microplastics, which are purposely introduced to textiles during production or which assemble on ocean-bound plastic debris, can also contain toxic compounds. The tiny aquatic organisms that consume them may harm them. According to reports, some researchers have discovered plastic microfibers in a range of often-eaten fish and shellfish. However, more study is required to fully comprehend the hazards that are brought by this pollutant [37, 38]. 2.9% of fish larvae suffered after consuming the microplastics [39]. The human food chain has recently been revealed to contain micro-debris in some food and drink products. In 2014, microplastics were found in German beer [40]; however, commercial table salt has been found to include microplastics in multiple investigations [41, 42]. The existence of microplastics (MPs) in potable water was first reported in 2017. There hasn't been conclusive evidence of microplastic translocation from prey to predator along the food chain. On the other hand, a study looked into the possible effects on people's health in people from various nations, including Finland, Italy, Japan, the Netherlands, Poland, Russia, the UK, and Austria. The immunological response of the digestive system may be impacted by microplastic particles, and they may also facilitate the spread of diseases and harmful substances. Microplastics can harm the digestive tract, blood circulation, liver, and lungs according to studies on animals [34].

4.5 Management of Microfiber Pollution

The durability, construction, material type, and chemical finish of a garment are just a few of the many variables that affect how much a garment sheds microfibers. Research on microfiber shedding is particularly difficult due to the complexity of textile production. Trapping microfibers before they leave the home may be one of the greatest options, as more and more microfibers are building up in our waterways every day.

4.5.1 Use Filter Bags/Fabric Filters

Many worldwide appliance manufacturers are working to produce built-in microfiber filters, although the majority of washing machines do not yet have them. In this regard, a few different kinds of external microfiber filters are now commercially available. These filters can be added to home washing machines and periodically disposed of the captured microfibers like how a dryer's lint screen is cleaned. Laundry bags and laundry balls are made specifically to collect microfibers while washing. An immediate and largely efficient strategy to stop the release of microfibers into waterways is by using these external washing machine filters and fiber-catching technology [37].

4.5.2 Microfibers Deterioration in the Aquatic Environment

Plastics remain in the aquatic ecosystem and decompose extremely slowly, particularly when they come into contact with environmental biological elements. Limited studies have assessed the degradation of textile fibers in aquatic environments [43]. Microfibers made of cotton and rayon ought to disintegrate in naturally occurring aquatic aerobic environments, however, polyester microfibers would remain for a long time, according to well-controlled aquatic biodegradation investigations [44].

Environmental elements that affect plastic degradation include internal parameters like molecular size, density, kind of polymer, and any extra chemicals, as well as external influences like agitation, UV exposure, and maybe biological activity. There are two types of degradation: biological degradation and non-biological degradation. In aquatic environments, plastic particles and fibers are prone to a variety of abiotic degradative processes, including hydrolytic degradation, photodegradation (UV radiation), and mechanical degradation [43, 45]. While it has been shown that MP particles in the marine surroundings carry a distinct biofilm known as the "plastisphere," it is doubtful that these bacteria would be responsible for the plastic material's deterioration. PET could be destroyed by a *Comamonas testosterone* F4 strain that was developed by Gong et al. and which is alkali-tolerant. This strain is known as F6 [30]. After as much as 50 h of incubation, the average size of the PET particles decreased from 7.3 to 1.58 to 2.63 m. *Zalerion maritimum*, a marine fungus, and other fungi have both been found to contribute to the breakdown of PET [46, 47].

4.5.3 Legislation on Microfiber Pollution

There are very few nations with legislation addressing microfiber pollution. France is the first nation in the world to pass legislation intended to prevent and control microfiber pollution from laundry. France's experience shows that legislation can effectively control microfiber contamination. France has controversial microfiber pollution legislation, yet it offers a workable solution to a serious ecological and environmental issue. Microfiber contamination is anticipated to be regulated on a much wider scale [48]. From February 2020, it is mandatory to equip all newly manufactured and sold washing machines with a microfiber filter by 2025 under French law [48].

4.5.4 Circular Economy Model for Mitigating Microfiber Pollution

In the past 20 years, only 15% of the 100 million tonnes of textiles consumed annually worldwide have been recycled [49]. In a linear economy, rapid fashion and "throwaway culture" are largely disposable, which contributes significantly to textile waste [50]. However, this causes an ever-growing environmental issue in addition to a massive loss of important resources. A novel strategy for cutting down on textile waste and minimizing microfiber pollution is the circular textile economy. In contrast to conventional methods like burning and disposal in landfills, recycling, and reuse can usually maintain fiber materials at their optimal value while minimizing environmental damage. It is more advantageous to reuse worn-out but still wearable clothing, such as cotton clothing and fashion made of synthetic fibers that don't age, through the sale of used items rather than recycling them [51, 52]. From the perspective of sustainable development, a circular textile economy will encourage and support changes to the textile value chain, particularly for several significant textile and apparel-producing nations in Asia and Africa. However, by emphasizing the ecological and environmental aspects, the circular textiles economy will promote improvements in the long-term end-of-life value of next-generation fibrous materials [48].

4.5.5 Technical Standards on Microfiber Pollution

Three technical standards on microplastic from end textile products [53–55] were approved in 2022 by the ISO/TC 38 subcommittees that deal with sampling, measuring material loss for microfibers from textile end-products by domestic washing method, and the qualitative and quantitative evaluation of microfiber from domestic washing. The development of novel fiber and yarn, as well as the design of clothing and fabrics, will be accelerated by the application of these evaluation criteria for microplastics from textile sources. A wide range of stakeholders, including producers of washing machines and supply chains for textiles and clothing, actively seek and support solutions throughout the whole lifecycle of fibrous materials. Future updates and releases of several technical standards and recommendations will be made available at different levels to control the production, reuse, recycling, disposal, and retrieval of textiles and clothing [48, 56].

4.5.6 Creating Awareness

By raising insight into the microfiber issue, microplastics can be avoided. Reducing microfiber loss during the washing process can be achieved by low-temperature washing, using a softener, washing a full load, and substituting liquid detergent for washing. Additional safeguards against microfiber contamination include:

- Promoting the reform of wastewater treatment
- Encourage regulation of washing machine upgrades
- Invest in biodegradable materials
- Make use of already existing remedies like the Cora Ball and Guppy Friend Washing Bag
- Chitosan and pectin-based finishing treatments
- Use a front-load washing machine and wash less frequently
- Use filter bags/fabric filters.

4.6 *Recent Research on Microfiber Pollution*

A study conducted on microplastics from a global viewpoint estimates that the quantity of microplastics in certain oceanic compartments will have doubled by 2030 [57]. Understanding the physical forms, applications, transit, and fragmentation of plastics into microplastics and nanoplastics is essential to developing solutions for this global issue. The primary sources of plastic debris and microplastics are the washing of textiles, at-sea losses, paint failure, tire wear, and the discharge of plastic waste onto land. After being consumed by organisms and dispersed by currents, plastics, microplastics, and biofuel accumulate and sink in surface waters where they eventually become part of the ocean's sedimentation. Apart from discharging hazardous substances into the surroundings, plastics serve as a foundation for biofilms that contain pathogens and non-native species. Microplastic quantity increases with decreasing fragment size. Particles smaller than 20 μm may cause cell membranes to rupture, increasing the risk. Consumption, metabolism, reproduction, and behavior can be negatively impacted by exposure to microplastics.

It is dangerous for people to consume contaminated water and shellfish. We spend more than 90% of our time indoors, which increases the risk of indoor microplastics due to the high concentration of polymeric materials. Enhancing methods for microplastic sampling and characterization, comprehending long-term behavior, additive bioavailability, and hazards to the well-being of organisms and ecosystems are a few of the scientific concerns. Reducing consumption, using more recycled plastic, and creating biodegradable additives and polymers are a few solutions.

The possibility that synthetic textile-derived microfibers could serve as a source of microplastics is investigated [58]. Research has been done on the microfibers that are produced during the washing process, the variables that influence the textiles' microfiber release, and the possibility that wastewater treatment plants (WWTPs) will release microfibers into the environment. A brief analysis of the possible harm that microfibers could cause to human health and marine and aquatic life was also done. According to studies, washing textiles and clothing release microfibers. Microfiber release varies on the fabric type and detergent used. It is challenging to identify clear trends with significant factors influencing the release of microfibers due to limited studies that are currently available and the often divergent results among investigations. Although current WWTPs are quite good at capturing microfibers, due to so many microfibers in the effluent, billions of fibers are generated into the environment every day. To make meaningful comparisons between studies, standardized methods, and procedures must be established.

A specific device's efficacy in monitoring and reducing anthropogenic pollution in drinkable water was examined [59]. The device was installed directly to the tap in both public and private buildings, and it was tested with drinkable water from several Slovenian cities. The outcomes confirmed the high level of microfiber contamination of potable water and showed how well the applied device removed both natural and synthetic microfibers from tap water.

One of the Southeast Asian cities with the highest levels of microfiber pollution is Kolkata. The microfiber pollution in wash effluents taken from various parts of Kolkata was taken into study. To remove enough microfibers and non-biodegradable materials from water samples, packed bed microfiltration (PBMF) was employed in this study in an economical and environmentally responsible manner. Additionally, effective parameters like packed bed height to diameter ratio (H/D) and mess size of the selected filtration unit were changed from 60 to 100 and 0.71 to 2.85, respectively, to better understand the efficacy of the method and to increase the possibility of using this substitute to alleviate concerns [60]. The current investigation shows that in an hour of operation at a flow rate of 60 L h1, the microfiltration efficiency of the proposed PBMF unit was increased to a maximum of 93.5% for sample A and 92.2% for sample D, respectively. Additionally, it was discovered that such a system may cost as much as $5 US annually.

The extensive analysis of papers across multiple disciplines revealed that a range of variables, such as the age of the textiles, the properties of the yarn and fiber, and laundry conditions, all affect the shedding of fibers. The primary laundry factors that influenced fiber shedding were the type of washing machine used, the volume of liquid used, higher temperatures, and the use of softeners. Variations in yarn twist, fiber type (filament/staple), production method, fabric structure, and specific density are some of the factors that affect the properties of the fiber and yarn. There is a research gap in the textile industry, as the review found that no work has been done on microfiber shedding using parameters specific to the textile industry [61].

Wet wipes and sanitary towels, two commonly flushed personal hygiene textile products, have not yet been demonstrated to positively correlate with microplastic fibers in aquatic environments. The amount of textiles used in unregulated sanitary and personal care products is not disclosed on the labels. White MP fibers can be found in the sediments near wastewater treatment plants (WWTPs), where they resemble the white fibers found in consumer sanitary products and sewage-related waste. Materials such as cellulose, polypropylene (PP), polyethylene terephthalate (PET), or a blend of PV and cellulose are used to make non-flushable wipes that are sold in stores [62]. Large volumes of cleaned-up sewage macro-debris, including sanitary towels and wet wipes, mixed with seaweed biomass, accumulated near the WWTP as a result of a combined sewer overflow. The microplastic fibers discovered in the garbage over ten months were similar to those discovered in the intertidal sediments near the WWTP. Nonetheless, the bulk of the fibers recovered from locations far from the WWTP were composed of ABS, PP, and polystyrene. The findings show that two underutilized environmental sources of white MP fibers are flushed sanitary napkins and wet wipes. Considering the nonwoven textile industry's global reach and projected expansion, more people need to be made aware of the dangers of disposing of sanitary products down the toilet, which can pollute the marine environment. This is preferable to using alternative land-based waste management methods [62].

Numerous important factors, such as the type of fabric, the knitting or weaving structure, the type and concentration of detergent, temperature, pH, spin speed, and the amount of time spent washing and drying, can influence the release of fiber fragments from clothing and household textiles [63]. When clothes and household textiles are washed, dried, and worn, tiny one to five-millimeter fiber fragments may be released, posing a new risk to the environment and public health. It is commonly known that aquatic life and the food chain are threatened by these sources of fiber. It describes how the release of fiber fragments is impacted by various mechanical and chemical textile treatments. The generation of fiber fragments during subsequent wet operations may be increased by any process, such as brushing, sanding, and bleaching, that weakens fibers and their interactions.

5 Conclusion and Future Prospects

A sizable volume of water-based effluent from the textile industry is produced and generated into the water bodies, where it becomes toxic with leftover chemicals like dyes and pigments. Agricultural and aquaculture practices are negatively impacted by some of the chemicals. In recent decades, industry and researchers have adopted a variety of modern technologies for wastewater effluent treatment. Several innovative methods have also been used to reduce water usage. Natural extracts, biomolecules, biomaterials, and biopolymers are gradually replacing some dangerous synthetic chemicals and auxiliaries. Surface modification, oxidation, laser etching, polymerization, printing, coating, water-free plasma, UV irradiation, laser, ozone, sandblasting, and other related technologies have all been used as pre-treatments, post-treatments, for in situ processing or post-polymerization, and to provide various functionalities regarding water, power, auxiliary, cost, and pollution. Nonetheless, there are some difficulties, including uniformity, fastness qualities, aging stability, and initial and ongoing costs. The study also covered probable microfiber pollution sources and how they harm our ecology. Microfibers are a significant concern for all organisms even if they enter water bodies in extremely small quantities. Strategies for managing microfiber contamination should place a high priority on efficient source reduction and the creation of affordable, effective remediation options. Although multistep treatment is necessary, wastewater treatment facilities should prioritize implementing cutting-edge ideas to address the pollution produced by sheds from synthetic clothing items. Improved waste management initiatives are needed, with a focus on recycling garbage, which could lessen the spread of microfibers into aquatic bodies. The development of new approaches to the problem should be the main focus of future research on this matter.

References

1. Samanta KK, Gayatri TN, Saxena S, Basak S, Chattopadhyay SK, Arputharaj A, Prasad V (2016) Hydrophobic functionalization of cellulosic substrates using atmospheric pressure plasma. Cellulose Chem Technol 50:745–754
2. Hasanbeigi A, Price L (2015) A technical review of emerging technologies for energy and water efficiency and pollution reduction in the textile industry. J Clean Prod 95:30–44
3. https://textilelearner.net/water-consumption-in-textile-processing-industry/
4. Shaikh H, Ayaz AM (2009) Water conservation in textile industry. Pak Text J 6:48–51
5. Saxena S, Raja ASM, Arputharaj A (2017) Challenges in sustainable wet processing of textiles. Textiles and clothing sustainability. Springer, Singapore, pp 43–79
6. Shukla SR (2007) Pollution abatement and waste minimisation in textile dyeing. In: Environmental aspects of textile dyeing. Woodhead, pp 116–148
7. Ahmed K, Das M, Islam MM, Akter MM, Islam MS, Al-Mansur MA (2011) Physico-chemical properties of tannery and textile effluents and surface water of River Buriganga and Karnatoli, Bangladesh. World Appl Sci J 12:152–159
8. Lakshmanan OS, Raghavendran G (2017) Low water-consumption technologies for textile production. In: Sustainable fibres and textiles. The textile institute book series, pp 243–265
9. Raichurkar P, Ramachandran M (2015) Effluent generated from the textile process industries. Int J Text Eng Proc 1:47–50
10. Sheshama M, Khatri H, Suthar M, Basak S, Ali W (2017) Carbohyd Polym 175:257–264
11. Teli MD, Pandit P, Basak S (2017) Coconut shell extract imparting multifunction properties to ligno-cellulosic material. J Indus Text 24:98–105
12. Teli MD, Pandit P (2017) Novel method of ecofriendly single bath dyeing and functional finishing of wool protein with coconut shell extract biomolecules. ACS Sustain Chem Eng 5:8323–8333
13. Basak S, Patil PG, Shaikh AJ, Samanta KK (2016) Green coconut shell extract and boric acid: new formulation for making thermally stable cellulosic paper. J Chem Technol Biotechnol 91:2871–2890
14. Rahman NAA, Tajuddin R, Tumin SM (2013) Optimization of natural dyeing using ultrasonic method and biomordant. Int J Chem Eng Appl 4:205–210
15. Basak S, Ali SW (2017) Leveraging flame retardant efficacy of the pomegranate rind extract, a novel biomolecule on lingo-cellulosic materials. Polym Degrad Stab 144:83–89
16. Basak S, Samanta KK, Chattopadhyay SK, Das S, Bhowmik M, Narkar R (2014) Fire retardant and mosquito repellent jute fabric treated with thio-urea. J Text Assoc 74:273–280
17. Al-Etaibi AM, El-Apasery MA (2021) Ultrasonic dyeing of polyester fabric with azo disperse dyes clubbed with pyridonones and its UV protection performance. Chemis 3:889–895
18. Gupta D, Basak S (2010) Surface functionalization of wool using 172 nm UV Excimer lamp. J Appl Polym Sci 117:3448–3453
19. Basak S, Samanta KK, Chattopadhyay SK, Pandit P, Maiti S (2016) Green fire retardant finishing and combined dyeing of proteinous wool fabric. Color Technol 132:135–143
20. Samanta KK, Pandit P, Samanta P, Basak P (2019) Water consumption in textile processing and sustainable approaches for its conservation. In: Water in textiles and fashion, pp 41–59

21. Das S, Bhowmick M, Chattopadhyay SK, Basak S (2015) Application of biomimicry in textiles. Curr Sci 109:893–901
22. Yang Y, Li S (2003) Cotton fabric inkjet printing with acid dyes. Text Res J 73:809–814
23. Hee U, Freche M, Kluge M, Provost J, Weiser J (2004) Ink jet interactions in ink jet printing-the role of pre-treatments, Textile ink jet printing—a review of ink jet printing of textiles. Society of Dyers and colourists Technical Monograph, 7, pp 44–56
24. Basak S, Samanta KK, Saxena S, Chattopadhyay SK, Parmar MS (2017) Self-extinguishable cellulosic textile from *Spinacia oleracea*. Ind J Fibre Text Res 42:215–222
25. Mishra S, Rath CC, Das AP (2019) Marine microfiber pollution: a review on present status and future challenges. Mar Pollut Bull 140:188–197
26. Laitala K, Klepp IG, Henry B (2018) Does use matter? Comparison of environmental impacts of clothing based on fiber type. Sustainability 10:1–25
27. Carr SA (2017) Sources and dispersive modes of micro-fibers in the environment. Integr Environ Assess Manag 13:466–469
28. Remy F, Collard F, Gilbert B, Compère P, Eppe G, Lepoint G (2015) When microplastic is not plastic: the ingestion of artificial cellulose fibers by macrofauna living in seagrass macrophytodetritus. Environ Sci Technol 49:11158–11166
29. Boucher J, Friot D (2017) Primary microplastics in the oceans: a global evaluation of sources. Gland, Switzerland
30. Gong J, Kong T, Li Y, Li Q, Li Z, Zhang J (2018) Biodegradation of microplastic derived from poly (ethylene terephthalate) with bacterial whole-cell biocatalysts. Polymers 10:1326
31. Liu J, Zhao X, Yang Y (2019) Beyond the definition of microfiber pollution is the debate over microfiber sources international concerns about since the official report about marine microplastics. AATCC Rev 19:49–53
32. Stanton T, Johnson M, Nathanail P, MacNaughtan W, Gomes RL (2019) Freshwater and airborne textile fibre populations are dominated by 'natural', not microplastic, fibres. Sci Total Environ 666:377–389
33. Henry B, Laitala K, Klepp IG (2019) Microfibres from apparel and home textiles: prospects for including microplastics in environmental sustainability assessment. Sci Total Environ 652:483–494
34. Bhatt P, Rani A (2022) Impact of microfibre pollution on the environment. https://www.fibre2fashion.com/industry-article/9428/impact-of-microfibre-pollution-on-the-environment
35. Chinglenthoiba Ch, Pukhrambam B, Chanu KTh, Devi Kh, Meitei NJ, Devika Y, Valiyaveettil S (2023) A review on microplastic pollution research in India. Region Stud Mar Sci:102777
36. Browne MA, Crump P, Niven SJ, Teuten E, Tonkin A, Galloway T, Thompson R (2011) Accumulation of microplastic on shorelines worldwide: sources and sinks. Environ Sci Technol 45:9175–9179
37. Pope M (2020) What you should know about microfiber pollution. https://www.neefusa.org/story/water/what-you-should-know-about-microfiber-pollution
38. Basak S (2023) India and microplastics: redefining India's one of the most crucial environmental problems. https://timesofindia.indiatimes.com/blogs/scientifically-trended/india-and-microplastics-redefining-indias-one-of-the-most-crucial-environmental-problems/
39. Steer M, Cole M, Thompson RC, Lindeque PK (2017) Microplastic ingestion in fish larvae in the western English Channel. Environ Poll 226:250–259
40. Liebezeit G, Liebezeit E (2014) Synthetic particles as contaminants in German beers. Food Addit Contam Part A 31:1574–1578
41. Yang D, Shi H, Li L, Li J, Jabeen K, Kolandhasamy P (2015) Microplastic pollution in table salts from China. Environ Sci Technol 49:13622–13627
42. Karami A, Golieskardi A, Keong Choo C, Larat V, Galloway TS, Salamatinia B (2017) The presence of microplastics in commercial salts from different countries. Sci Rep 7:1–11
43. Zambrano MC, Pawlak JJ, Daystar J, Ankeny M, Cheng JJ, Venditti RA (2019) Microfibers generated from the laundering of cotton, rayon and polyester based fabrics and their aquatic biodegradation. Mar Poll Bull 142:394–407

44. Palacios-Mateo C, Van der Meer Y, Seide G (2021) Analysis of the polyester clothing value chain to identify key intervention points for sustainability. Environ Sci Europe 33(1):2
45. Gewert B, Plassmann M, Sandblom O, MacLeod M (2018) Identification of chain scission products released to water by plastic exposed to ultraviolet light. Environ Sci Technol Lett 5:272–276
46. Sánchez C (2020) Fungal potential for the degradation of petroleum-based polymers: an overview of macro-and microplastics biodegradation. Biotechnol Adv 40:107501
47. Gaylarde C, Baptista-Neto JA, da Fonseca EM (2021) Plastic microfibre pollution: how important is clothes' laundering? Heliyon 7(5)
48. Liu J, Liu Q, An L, Wang M, Yang Q, Zhu B, Xu Y (2022) Microfiber pollution in the earth system. Rev Environ Contamina Toxico 260:13
49. Shirvanimoghaddam K, Motamed B, Ramakrishna S, Naebe M (2020) Death by waste: fashion and textile circular economy case. Sci Tot Environ 718:137317
50. Bucknall DG (2020) Plastics as a materials system in a circular economy. Philos Trans R Soc A 378:20190268
51. Cao Y, Qu Y, Guo L (2022) Identifying critical eco-innovation practices in circular supply chain management: evidence from the textile and clothing industry. Inter J Logis Res Appl:1–22
52. Sandin G, Peters GM (2018) Environmental impact of textile reuse and recycling–a review. J Clean Prod 184:353–365
53. ISO (2021a) Textiles and textile products—Microplastics from textile sources—Part 1: Determination of material loss from fabrics during washing. https://www.iso.org/standard/82238.html
54. ISO (2021b) Textiles and textile products—Microplastics from textile sources—Part 2: Qualitative and quantitative evaluation of microplastics. https://www.iso.org/standard/80011.html
55. ISO (2021c) Textiles and textile products—microplastics from textile sources—Part 3: Measurement of collected material mass released from textile end products by domestic washing method. https://www.iso.org/standard/81035.html
56. Simon N, Raubenheimer K, Urho N, Unger S, Azoulay D, Farrelly T, Weiand L (2021) A binding global agreement to address the life cycle of plastics. Sci 373:43–47
57. Hale RC, Seeley ME, La Guardia MJ, Mai L, Zeng EY (2020) A global perspective on microplastics. J Geophy Res Oceans 125:14719
58. Acharya S, Rumi SS, Hu Y, Abidi N (2021) Microfibers from synthetic textiles as a major source of microplastics in the environment: a review. Text Res J 91:2136–2156
59. Volgare M, Avolio R, Castaldo R, Errico E, El Khiar H, Gentile G, Sinjur A, Susni D, Znidarsic A, Cocca M (2022) Microfiber contamination in potable water: detection and mitigation using a filtering device. Microplas. 1:322–333
60. Mondal I, Ghosh D, Biswas PK (2023) Cost-effective remedial to microfiber pollution from wash effluent in Kolkata and Ranaghat. Chemosphere 313:137548
61. Rathinamoorthy R, Raja BS (2021) A review of the current status of microfiber pollution research in textiles. Int J Cloth Sci Technol 33:364–387
62. Briain OÓ, Mendes ARM, McCarron S, Healy MG, Morrison L (2020) The role of wet wipes and sanitary towels as a source of white microplastic fibres in the marine environment. Water Res 182:116021
63. Periyasamy AP, Tehrani-Bagha A (2022) A review on microplastic emission from textile materials and its reduction techniques. Polym Degrad Stab 199:109901

Eco-friendly Dyeing Approach: Natural Dyeing—A Need of the Hour

Arpana Kamboj, Meenakshi Tamta, Pooja Kundal, and Bhawna Soun

Abstract The growing need for human and environmental protection in various fields and the world of fashion leads to the fact that the finishing of textiles for the environment gives ease of use. Colours have a unique place in our lives, and they are used to beautify not only the environment but also the textiles. In the realm of textile dyes, there has been a rise in the need for sanitary, fresh, and clean textiles. This has led to a rise in the use of natural dyes. Since natural dyes offer remarkably unique, calming colours and tones in contrast to synthetic dyes, people are now well aware of the many benefits of using them. In addition, natural colours are more sustainable and biodegradable than artificial ones. Natural dyes are renowned for their exceptional stainability. Different types of natural dyes have been reported over the past few decades, with variations in their synthesis and application methods on various tactile. This chapter reviews various research attempts to infuse natural dyes into texiles. The influence of environmentally friendly natural dyes on humans, and the need for natural dyes as antimicrobial, antifungal and UV absorbers in textile finishing are well covered.

Keywords Natural · Antimicrobial · Mordant · Dyes · Synthetic · Extraction · Textile

A. Kamboj (✉) · M. Tamta
Department of Fashion and Textile Design, Swami Vivekanand Subharti University, Meerut, Uttar Pradesh 250005, India
e-mail: arpanakamboj@gmail.com

P. Kundal
Department of Krishi Vigyan Kendra, Muzaffarnagar-II, Sardar Vallabhbhai Patel University of Agriculture and Technology, Meerut, Uttar Pradesh, India

B. Soun
Department of Apparel and Textile Science, Punjab Agricultural University, Ludhiana, Punjab 141027, India

Sadhna et al. (eds.), *Climate Action Through Eco-Friendly Textiles*, SDGs and Textiles,
https://doi.org/10.1007/978-981-99-9856-2_7

1 Introduction

An eco-friendly dyeing approach involves using natural and sustainable materials and processes to colour texiles while minimizing the impact on surroundings. Natural colouring agents are derived from naturally occurring substances and are usually recyclable, which makes them a greener choice compared to artificial colours. Natural dyes, which come from plants, minerals, and animals, can create a wide range of colours. Because natural dyes are sustainable, renewable, biodegradable, and friendly to the environment, their use has been revived. Natural dyes are derived from a variety of plant sources [1].

In contrast to synthetic colours, natural dyes produce extremely rare, glossy, calming, and delicate shades. When natural dyes replace synthetic colours made from petroleum-based fossil fuels, they can produce carbon credits. Due to the numerous advantages that synthetic dyes have over natural dyes, such as colour fastness, good shade reproducibility, colour brilliance, ease of use, cost savings, and the availability of pure synthetic colours of all kinds and classes, the majority of textile manufacturers and dyers have switched to using synthetic colourants [2].

Consumer concerns regarding the sustainability and environmental friendliness of the products they use have increased in the current environment, and natural dyes are starting to become somewhat more popular. Natural dyes were primarily derived from plants, animals, and minerals. To create different colours and their combinations, plants' entire range of parts—including leaves, roots, fruits, flowers, bark, wood, and seeds—was utilised [3].

Most pigments used in the process of extracting dyes are considered medicinal, though some have shown promising antimicrobial activity recently [4]. The tannins contained in *Punica granatum* and natural dyes have antibacterial properties. Gershon and Shank discovered that plant colours with a lot of naphthoquinones, like lawsone from henna, juglone from walnut, and lapachol from alkanet, have antibacterial and antifungal qualities [5].

The topic of the current study is environmentally friendly dyes. These dyes are incredibly helpful and give the textile industry a sustainable and eco-friendly alternative. These dyes don't release dangerous chemicals into the environment. Natural dyes are also incredibly beneficial to health. These are not harmful to human health and are safer than synthetic dyes. Numerous synthetic dyes are recognised to contain hazardous chemicals that can lead to health complications like skin irritation and respiratory problems. Because these dyes are healthy and environmentally friendly, they can be very helpful in the production of clothing for the elderly and infants, protecting them from various infections. These dyes will also be useful in home textiles such as bed linen, rug, and table linen, which serve as major carriers of common infections.

Natural dyes generate an extensive variety of distinct and beautiful colours that are unable to be replicated by synthetic dyes. Each plant, insect, or mineral source produces its distinct colour, resulting in a rich diversity of hues.

Overall, the importance of natural dyes lies in their sustainability, health benefits, unique colours, cultural heritage, and economic benefits. By using natural dyes, we can promote a more sustainable and environmentally friendly approach to textile production while also preserving cultural traditions and supporting local economies.

2 Eco-friendly Dyes

Natural sources such as fungi, minerals, plants, and animals are the source of natural dyes. Natural dyes come from plants, animals, minerals, fungi, and other natural sources. They have played a crucial role in many cultures and customs since they were first used thousands of years ago to colour fibres, yarn, and fabrics. These dyes have unique and varied colours, and because they are eco-friendly, biodegradable, and non-toxic, they are often chosen over synthetic dyes. Several instances of natural dyes are indigo, madder, cochineal, and henna. It is observed that due to worldwide environmental consciousness, people were growing their enthusiasm for the resurgence of natural dyes on organic fabrics [6].

The chemical substances that plant naturally make as part of their defence mechanisms and used for defense against numerous diseases are known as phytochemicals. Alkaloids, steroids, phenolic compounds, flavonoids, anthraquinones, and tannins are significant phytochemical groups that are found in numerous plant extracts and are responsible for their natural therapeutic qualities [7]. Plank tannin has pharmacological properties and serves a broad variety of functions, such as antibacterial, antiviral, antimicrobial, and anti-tumour properties. Tannins are compounds that can give a product its Natural colour and are vital components of dyes. Alkaloids are a very powerful pesticide and have physiological effects that are utilized in nicotine sulphate, a tobacco industry by-product. In addition to possessing antifungal and antibacterial effects, saponins have medical uses as an emulsifier and expectorant. Glycoside is frequently found in natural medicines. Steroids can be found in abundance in the leaves, stems, bark, and roots. They are also used in nutrition, medicinal herbs, skin care products, and medicine due to their potent biological activity. Terpenoids are employed to prevent competing plants from sprouting and growing, and their aroma in fruits and flowers draws insects that disperse pollen or seeds.

Anthraquinones are thought to be involved in natural plant disease resistance and frequently exhibit antibacterial effects. The compounds known as flavonoids are those that can impart colour [8, 9]. The colour and antibacterial qualities of textiles are also due to phytochemicals [10].

3 Importance of Eco-friendly Natural Dyes

Natural dyes have the power to add distinctive, visually appealing qualities to fabric or clothing while also giving an ecologically friendly product moral weight. These enhance the aesthetic and industrial aspects of the textile production process even further. The natural dye industries have a financial incentive to produce and distribute dyes in an environmentally friendly manner in addition to being a plentiful and diverse source of dyes [11, 12]. Natural dyes are widely used in a variety of industries, including confectionery, food processing, cosmetics, pharmaceuticals, leather products, papers, paint and ink and so on.

The market for antibacterial textiles is currently seeing a rise in demand. Because a considerable number of phytochemicals (Table 1) are present, and some natural dyes also have antimicrobial properties [13]. It has been demonstrated that several natural dyes possess analgesic, antiviral, antibacterial, antifungal, antileprotic, and anti-inflammatory qualities. Bioactive phytochemical components, which have physiological effects on human bodies, such as flavonoids, alkaloids, and other essential oil tannins, terpenoids, saponins, phenolic compounds, and so forth, are the source of plants' medicinal potential [14].

4 Classification of Eco-friendly Dyes

Natural dyes are substances that are extracted without the use of chemicals from a variety of natural sources, including insects, plants, animals, and minerals [16].

4.1 Plant-Based Dyes

Plant components like leaves, roots, branches, stems, wood shavings, fruits, flowers, hulls, and husks are the source of these hues [17, 18]. Natural dye sources can also be found in other organic materials like lichens and bacteria. These are environmentally beneficial substitutes for synthetic dyes due to their biodegradable, renewable, and non-toxic nature. The following are a few examples of plant-based dyes: henna, hibiscus, marigold, orange peel, and turmeric.

Reddish-orange in hue, henna has been used for ages to colour clothes, hair, and skin. The application and extraction of henna dye in textile materials have been the subject of extensive research, and methods for standardising and streamlining the dyeing process have been found. Because henna contains polar molecules, which give it an acidic tendency, using it in the textile dyeing process is encouraged [19]. A significant source of material for natural dyeing is hibiscus. It is a member of the Malvaceae family. These flowers' aqueous extracts have demonstrated good fastness qualities. The dye has been found to possess a variety of uses, including the

Table 1 Different phytochemicals present in natural dyes [15]

Berberine	Ellagic acid
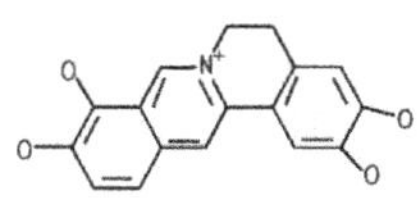	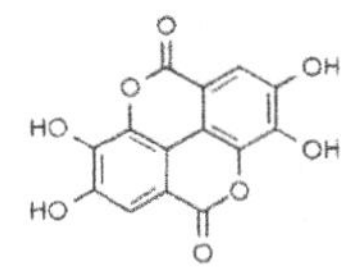
Juglone	Gallic
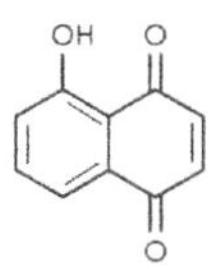	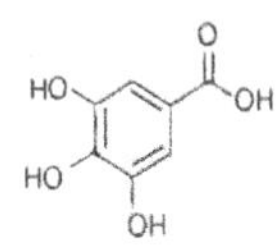
Flavonoid	Lawsone
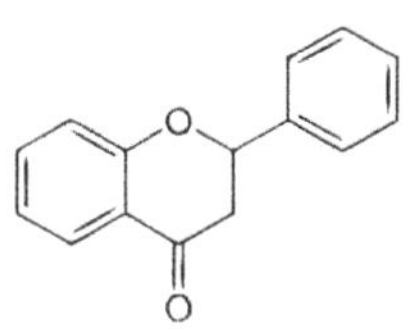	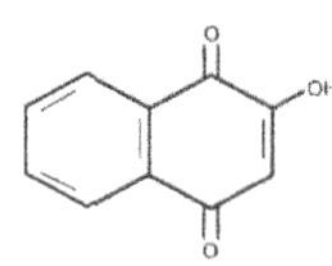
Curcumin	Phytol
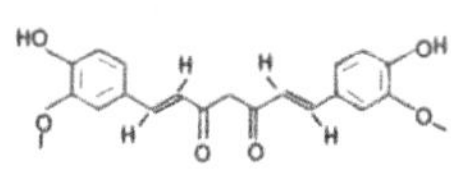	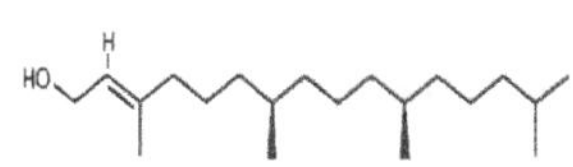
Tannic acid	
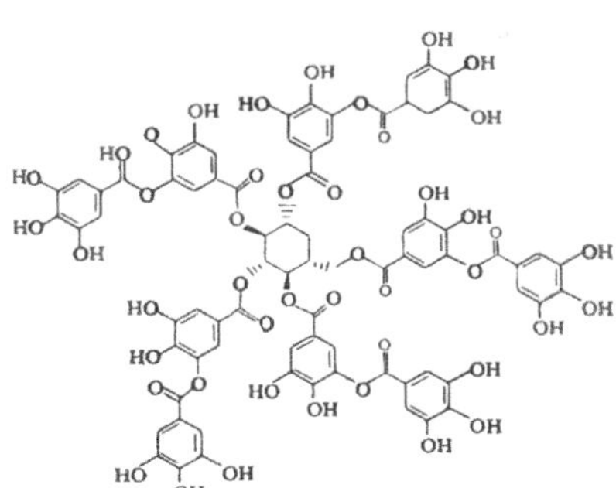	

commercial colouring of cotton, silk, and wool yarn for the carpet industry and clothing manufacturing [20]. Orange peel is a readily available, inexpensive, and plentiful agricultural by-product. The textile materials were dyed with orange peel using environmentally friendly iron and aluminium mordants. The colour fastness to laundering, rubbing fastness, and light fastness of the textile material dyed with peel orange is good [21]. The marigold plant can be used to dye wool with a yellow colour. The wool fabrics were treated to different concentrations of ammonia solution after being pre-mordanted with alum and coloured with marigold. Although the colour changes after brushing with conventional detergent and ammonia treatment do not affect washing fastness, the samples' light fastness has decreased [22]. Turmeric dye is derived from the plant's root. The root is dried, ground into a powder, and the dye was extracted by boiling it in water. It works well for dying silk, wool, and cotton. The property of colour fastness is enhanced by mordant. Turmeric dye produces a vibrant shade of yellow when dyed. There are a lot of curcuminoids in this dye. Turmeric contains a colouring substance called curcumin [23].

4.2 Animal-Based Dyes

Additionally, naturally occurring sources of dyes include insects, molluscs and crustaceans. They are known for their vivid and long-lasting colours. The greatest red dyes used back then were of animal origin and came in a variety of colours of red and purple. The main source of natural colours is insect secretions and dried insect bodies [24]. Some examples of animal-based dyes include cochineal, Lac insect (*Laccifer Lacca* Kerr) and shellac.

Cochineals, kermes and lac are sold in the market as small, dark-coloured grains that, when combined with hot water, produce a carmine-coloured solution that contains a significant amount of the colouring agent carminic acid. The dried carcasses of many insects that reside on a species of oak make up kermes. The dye is extracted from the body of the coccus laccac, a tiny bug. Tyrian purple is a highly sought, ancient molluscan dye that is made from the juice of specific kinds of snails that are found in the waters of the Mediterranean Sea. It is extremely expensive because 12,000 animals were required to produce only one gram of the colour. Another significant and ancient animal source, lac, also provides a red colour. Lac is a red dye that is water-soluble and made from the venomous secretion of the small bug Leacciferlacca. After the dying process, it turns a scarlet to crimson red colour [23].

4.3 Mineral-Based Dyes

These dyes are derived from minerals, such as iron, copper, and aluminium. They are known for their earthy and muted colours and can be found in Natural pigments like ochre and sienna.

Iron-based dyes produce shades of grey, black and brown. Iron oxide can be found in many Natural sources, such as rust, clay and hematite. Green and blue tones are produced by a dye made of copper. A particular type of mordant used to keep the dye on the fabric is copper sulphate. Dyes based on tin yield orange and yellow hues. Tin is frequently added to natural dyes as a mordant to help make the colours more vibrant. Blue and purple hues are produced by aluminium dyes. Alum is commonly used to mordanize natural colours. In general, mineral-based natural dyes are less likely to fade over time than plan-based natural dyes because they are more light fast and colour fast. However, to get the right colour and durability, they might need more intricate processing and mordanting methods. All things considered, natural dyes made of minerals provide a distinctive and environmentally friendly substitute for synthetic dyes, in addition to offering a variety of stunning and durable hues.

4.4 Fungi-Based Dyes

These dyes come from lichens and mushrooms, among other types of fungi. They can create distinctive shades that are challenging to duplicate with synthetic dyes and are renowned for their vivid and varied colour palette. Lichen green and Tyrian purple are two hues made from fungi.

Microorganism-derived pigments are superior to plant- or animal-derived pigments in many ways, including yield, labour costs, supply sustainability, stability, and ease of further processing [25]. Microalgae and fungi are two additional microbial sources that produce an astonishing array of water-soluble biopigments with a broad range of ecological applications [26]. Basidiomycetous fungi, which were used by ancient societies for dyeing wool and silk fabric, are being used to produce biopigments [27]. Many colours made by filamentous fungi are available on the market, such as anka flavin and cantha-xanthin and Ar-pink red pigment (Natural redTM) from the strain *Penicillium oxalicum var. arme-niaca* [28].

In a field that is evolving quickly, natural colours have surpassed synthetic ones in popularity, but research into fungal-based pigments still has to be done to realize their promise as a source of industrial pigment in the future.

4.5 *Fruit-Based Dyes*

Natural fruit dyes are made from a variety of fruits and vegetables, such as onion skins, pomegranates, avocado pits, and berries. These natural resources can produce a vast array of beautiful and vibrant colours and are more sustainable and kind to the environment than synthetic dyes.

Berries like raspberries, blackberries, and blueberries develop pink, purple, and blue hues. Berries are often boiled in water to extract these colours, and the resulting liquid is then used as a dye bath. You can add shades of green and yellow with pomegranate peels. After the skins are boiled in water to extract the dye, the cloth is immersed in the dye bath. The colour of avocado pits can vary from peach to pink to light beige. The fabric is coloured with a liquid that is obtained from the dye by boiling the pits in water. You can use onion peels to get shades of brown, orange, and yellow. The skins are steamed in water to extract the dye, and the resulting liquid is used to colour the fabric.

Fruit-based natural dyes are generally safe and non-toxic and can be easily sourced from local farmers' markets or grocery stores. They can create a range of lovely and distinctive colours and offer a sustainable and eco-friendly substitute for synthetic dyes [29].

5 Extraction Processes for Natural Dyes

Natural colouring agents have a complex structure rather than being a single chemical entity. Natural plant dye extraction is a very complex and precise process because the plant matrix contains both dye and non-dye constituents. The following are the various methods for extracting colouring materials:

5.1. **Aqueous Extraction**: The traditional method for removing dyes from plants and other materials is called aqueous extraction. The dye-containing material is either reduced to minute fragments or powdered and screened to increase the extraction process' efficiency. The material is soaked for a predetermined amount of time in earthen, metal, wooden, and stainless steel vessels in order to loosen the structure of the plant's cells. After soaking, bring the water to a boil to obtain a filtered solution free of any remaining non-dye. To extract as much dye as possible, the boiling and filtering steps are repeated. Larger scale pure colour powder extraction uses stainless steel vessels; longer boiling times of the solution can shorten the time the ingredients need to soak in water. Centrifuges are typically used to separate microbiological material that is left over. Trickling filters can be used to increase the solubility of the purified natural dye and decrease minute particles of plant material [29].

Since only the colour components that are soluble in water are removed, this extraction method has several disadvantages, including a lengthy reaction time, a high water requirement, the use of high temperatures, and a low dye yield. Many dyes have a low solubility in water [2].

5.2. **Acid and Alkali Extraction Process**: Weak acidic or alkaline solutions are effective in removing most dyes that contain glycosides. Glycosides are broken down by acid and alkali, which improves extraction and produces a good colourant matrix. When it comes to stopping oxidation, acid extraction works best with dyes that contain flavone groups, while alkali extraction works best with dyes that contain phenolic compounds to increase dye output. Glycosides can be eliminated by using diluted alkaline or acidic solutions because they come in a variety of colours. The hydrolysis of glycosides is aided by the addition of acid or alkali, increasing the yield and extraction of colourant components. The dye is extracted from Tesu (Buteamonosperma) flower petals using an acid hydrolysis process. Some flavone dyes are exposed to acidic water to prevent oxidative degradation. Because phenolic groups are dissolved in alkali, alkaline extraction is suitable for dyes containing phenolic groups because it increases colour yield. By using acids, dyes can later be precipitated. Using this method, dye from annatto seeds can be removed.

5.3. **Ultrasonic and Microwave Extraction**: This process reduces the amount of time, solvent, and temperature needed while increasing extraction efficiency. Ultrasound treatment of plant material causes tiny bubbles, sometimes known as cavitations, to form in the liquid. High temperature and pressure caused the bubbles to burst, which served to improve the effectiveness of the extraction process. In the microwave extraction method, plant material is processed in the presence of microwave energy sources with a minimal amount of solvent. The use of a microwave helps to accomplish this operation more quickly and with better outcomes.

The microwave extraction process uses the least amount of solvent when processing organic materials in the presence of microwave energy sources. Because of how quickly and efficiently the microwave speeds up the process, the extraction can be finished. The annatto colourant was eliminated using the microwave energy [30]. Their team had previously described the butterfly pea's blue pigment being extracted with the help of a microwave. Because they require less energy due to lower extraction temperatures, fewer solvents, and shorter extraction times, ultrasound extractions can be regarded as green processes.

5.4. **Fermentation**: After being freshly collected, indigo leaves and twigs are soaked in hot water (about 32 °C). The indi-muslin enzyme, which is also found in foliage, converts the colourless indigo-containing glucoside indican in the leaves into glucose and indoxyl as fermentation progresses. The yellow liquid containing indoxyl is transferred to beating vats after ten to fifteen hours of fermentation. Here, oxygen oxidises the indoxyl to produce blue, insoluble indigotin that sinks to the bottom. After the extra water is extracted, it is collected, cleaned, and shaped into cakes. Another method for extracting indigo from other materials, such as wood, is fermentation. This method can also be used to extract other colourants, such as annatto.

5.5. **Enzymatic Extraction**: The enzymes like cellulase, amylase, and pectinase is used by some researchers to extract the colour by breaking down the fibrous matrix and making it safer to extract colour molecules. This is because the plant tissues contain pectins, starches, and cellulose, which act as binding agents. This technique may be helpful for colouring plant parts that are difficult to reach, such as the roots and bark.
5.6. **Solvent Extraction**: Organic solvents like acetone, petroleum-based ether, methanol, and chloroform, as well as solvent mixtures like ethanol, water, and alcohol, are used to extract natural colouring agents. The water extraction method can be used to extract materials from plant resources that are soluble or insoluble. In comparison to the water-soluble method, a higher variety of chemicals and colouring ingredients can be extracted, increasing the extraction yield. When mixed with an acid or an alkali, alcoholic solvents can also be used to aid in the hydrolysis of glycosides and the release of colourants.

6 Processing of Natural Dyes

While some standard procedures are often followed, the extraction process for natural dyes can vary depending on the specific source of the dye. Extraction of natural dyes can be a tedious procedure requiring meticulous attention to detail [31]. An overview of the steps involved in extracting natural dyes is provided below.

6.1. **Collection of the dye source**: The dye's source—a plant, mineral, or animal—is gathered and made ready for extraction.
6.2. **Preparation of the dye source**: The dye may need to be ground, chopped, or soaked in water to release its colour, depending on where it came from.
6.3. **Extraction of the dye**: Alcohol, water, or other solvents are used to remove the dye from its sources. The dye source may be soaked in a solvent for a period of time or simmered in water as part of the extraction procedure.
6.4. **Straining the dye**: The dye is strained to get rid of any solid particles or contaminants after it has been extracted.
6.5. **The concentration of the dye**: One way to concentrate the extracted dye is to simmer it over low heat until the desired intensity is achieved.
6.6. **Mordanting the fabric**: To help the dye stick to the fibres, the fabric must be mordanted before being dyed. Fabrics can be dyed using mordants like tannins, iron, or alum.
6.7. **Dyeing the fabric**: After soaking in the dye bath, the mordanted cloth is simmered for a while to get the right colour.
6.8. **Rinsing and finishing**: The cloth may be treated with a finishing agent to enhance its texture or durability after dying and is then carefully cleaned to remove any remaining dye.

7 Different Types of Mordents

The word "mordant" comes from the Latin word "mordere," which means "to bite." Pigments and fibres are bound together by a chemical compound that forms a complex integration between the two. It increases the dye's colour fastness and intensifies its colour [32]. In the dying process, a variety of natural mordents can be employed, such as:

7.1. **Alum**: Alum, or potassium aluminium sulphate, is an earthly natural mordant that is widely used. It creates incredibly vivid and bright colours and binds well with a variety of natural dyes. Alum is a popular option for natural dyeing because it is also relatively safe and easy to work with. Owing to the Mediterranean region's rich alum deposits, alum and iron were widely used as mordants in Egypt, India, and Assyria [33].

7.2. **Iron**: When combined with specific dyes, ferrous sulphate, or iron, is a natural mordant that can create muted, darker colours. Additionally, iron can aid in improving the colour fastness of natural dyes.

7.3. **Tannins**: Natural substances called tannins can be found in many plant materials, such as pomegranate rinds, tea, and oak galls. They can create soft, muted colours and can be used as a natural mordant to help bind with some types of dyes.

7.4. **Copper**: Green tones can be produced by combining copper, a natural mordant, with specific dyes. Use caution when using it as large amounts can be toxic.

7.5. **Cream of Tartar**: A natural material called potassium bitartrate, or cream of tartar, can be used as a mordant to aid in the binding of some dyes. It also has the ability to change the pH of the dye bath.

7.6. **Soy milk**: In Japan, soy milk is a common natural mordant used to prepare fabric for indigo dyeing. It facilitates the development of a strong bond between the pigment and the fabric's fibres, resulting in a vibrant, long-lasting colour.

8 Advantages of Eco-friendly Dyes for the Environment and Human Beings

Eco-friendly dyes offer several advantages for both the environment and human beings, including:

8.1. **Reduced environmental impact**: Since these dyes are derived from nature, they don't contain any components that are harmful to the environment. Customers find natural dyes appealing since they are environmentally friendly, sustainable and biodegradable [34].

8.2. **Availability**: Natural dyes are widely available in nature. They can therefore be conveniently gathered and used as required. These colours are widely available in nature and are easily obtainable from the surroundings [35].

8.3. **Health benefits**: Eco-friendly dyes are safer for humans to use and wear, as they do not contain toxic chemicals that can cause skin irritation, allergies, or other health problems. This is especially important for people who are allergic or sensitive to synthetic dyes. Numerous experts have suggested and documented the value of natural colours for medicine and antimicrobial purposes [36]. Natural colours are healthier options because they don't include any harmful chemicals for your health [37]. For example, turmeric's rhizome contains a yellow colour that has been traditionally utilized as an anti-inflammatory medication in medicine [38].

8.4. **U.V. protective fabrics**: Sunburns, tanning, premature skin burns, and skin ageing are all prevented by wearing ultraviolet-protected clothing. Wool, silk, and cotton may have higher UPF (ultraviolet protection factor) values when treated with metallic mordants. When natural orange peel extract dye was used to dye wool fabrics, the UPF property significantly increased [39].

8.5. **Cheap**: Natural dyes are inexpensive, accessible, and safe to use for dying. Compared to synthetic dyes, some natural colours can be cheaper [40].

8.6. **Sustainability**: Eco-friendly colours are made from renewable sources, such as plants, and are often produced using environmentally sustainable methods that conserve water and energy. They are biodegradable, minimizing waste in landfills and the oceans [30].

8.7. **Insect resistant**: Moths and fungi have an impact on wool and cellulose materials at high temperatures. Wool is coloured using dyes based on anthraquinone. Materials that are insect-resistant and insect-repellent can be made using natural dyes like cochineal, indigo, and madder [41].

8.8. **Non-Toxic**: Natural dyes are safe, non-toxic, and do not affect skin allergies. They are generally more beneficial to the human body than synthetic dyes [42].

8.9. **Support for local communities**: Not only do eco-friendly dyes offer economic benefits and safeguard cultural heritage but they are also made using traditional methods and often originate from nearby communities. Natural dyes are not only abundant and diverse sources of colourants, but they can also be an eco-friendly way to make money by harvesting and selling specific dye plants [11].

9 Applications of Natural Dyes Other Than Textile Dyeing

Natural dyes are being utilized for a wide range of applications. They are applied for various purposes, such as pH indicators, food and cosmetic colourants, cosmetic healing additives, functional finishing (antimicrobial, antifeedant, deodorizing or UV protection), and textile dyeing and finishing [43]. Food, leather and cosmetics all contain natural dyes. It is generally safe for people's health and safe for skin contact because some of its ingredients are anti-allergens [44]. Because they have fewer adverse effects than artificial dyes, natural dyes are becoming more and more common in the cosmetics industry. However, they may also provide additional benefits like UV protection, skin hydration and anti-ageing [45]. Antibacterial coatings may assist in reducing the dissemination of disease, minimize the risk of infection from injuries and maintain fabric quality [46].

10 Industrial Dyeing Procedures and Their Adverse Effects

In the more than 4000 years that the textile dyeing industry has existed, the industrial production and use of synthetic dyestuffs has expanded into a massive industry. However, all living things suffer when artificial dyes are used [47]. The following are some of the negative environmental effects of industrial dyeing procedures.

10.1. **Water waste**: Up to 125 L of water are needed for each kilogramme of cotton fibre during the dyeing process. With many Western brands outsourcing their workforce and having little control over supply chain management, this places a great deal of strain on the world's aquatic resources, particularly those of developing nations. According to the Ellen MacArthur Foundation, the fashion industry uses about 93 billion cubic metres of water annually, or enough to fill 37 million Olympic-sized swimming pools [48].

10.2. **Pollution**: The presence of naphthol, vat dyestuffs, nitrates, acetic acid, soaping chemicals, enzymatic substrates, chromium-based materials, heavy metals and other dyeing auxiliaries makes the wastewater from textile dyeing extremely toxic. In addition to endangering human health, synthetic dyes devastate the environment and ecosystems. They generate hazardous chemical waste that harms rivers and other water sources. The death of soil microorganisms by harmful dyes has an impact on agricultural productivity. Particularly harmful to the ecosystem are azo dyes [49].

10.3. **Health hazards**: Textile dyes are extremely toxic and may cause cancer, as are many other industrial pollutants. The air, drinking water and agricultural soil are all contaminated by toxic chemicals found in fabric treatments and dyes. Both human health and ecosystems are severely harmed by them [48].

The textile dyeing industry has looked into innovations, eco-friendly remediation techniques, and alternative ecosystems to lessen the negative environmental effects of industrial dyeing procedures. A variety of ecologies have been investigated, including waterless dyeing with supercritical carbon dioxide assistance, natural mordants and colours. Dyeing clothing exclusively with bio-based dyes is a more environmentally friendly way to lessen the impact of fashion dyes on the environment.

11 Natural Dyeing—Need in the Global Scenario

Natural dyeing is a viable and ecologically conscious substitute for synthetic dyeing, which has long been associated with negative effects on the environment and human health. The importance of natural dyeing is becoming more and more clear in the modern, global context, for several compelling reasons.

One of the most obvious advantages of natural dyeing is that it has less of an impact on the environment. Conventional synthetic dyes emit a torrent of dangerous chemicals into the atmosphere, causing irreversible damage to the environment. By switching to natural dyes, we can directly oppose these dangerous substances and mitigate the harm they cause to the environment. Using natural dyeing instead of synthetic dyeing results in a significant reduction of the toxic waste that is typically produced by textile dyeing processes. This sustainable deed begins to make up for the damage. The lower environmental impact of natural dyeing is one of its most striking benefits.

Conventional synthetic dyes damage the environment irreversibly by releasing a torrent of hazardous chemicals into the atmosphere. Adopting natural dyes allows us to directly oppose these harmful substances and lessen their negative effects on the environment. The toxic waste that is normally produced by textile dyeing processes is significantly reduced when natural dyeing is used instead of synthetic dyeing. This sustainable action starts to offset the harm caused by the traditional textile industry, which has long been associated with environmental issues.

An additional essential component of natural dyeing is sustainability. Natural dye plants can grow in otherwise unutilized wastelands, which is an effective use of resources that were previously disregarded. To further reduce waste and increase efficiency, the waste products made during the natural dyeing process can be recycled into organic fertilisers. The process of making natural dyes is also more labour-intensive because it uses human abilities rather than machinery. This shift creates job opportunities and strengthens the bond between employees and their jobs. Our dependency on fossil fuels is decreased by the general move towards natural dyes, which is an important step towards a more sustainable future.

A vital component of natural dyeing is its health benefits, which stand in stark contrast to the hazards that come with using synthetic dyes. Hazardous substances like ammonia, chlorine, and heavy metals are frequently found in synthetic dyes, and they can penetrate our skin and harm our health. The seriousness of the problem is highlighted by the World Bank's concerning estimate that 72 hazardous chemicals are released into the water during the textile dyeing process, with over 30 of them potentially harmful to human health directly or indirectly. In contrast, natural dyes are devoid of hazardous chemicals and safe for human skin, making them a safer and healthier choice for both consumers and textile workers.

Even with all of the benefits associated with natural dyeing, the demand for this eco-friendly colourant is still relatively small worldwide—it makes up only 0.1 million tonnes of colourants annually, or 1% of all colourants consumed synthetically. However, there are currently large-scale research projects underway in the world aimed at refining and enhancing natural dye manufacturing processes for commercial use. The natural dyeing industry is still evolving, which is a sign of its potential and the increasing awareness of sustainable practices around the world. Even though there is still a long way to go, natural dyeing is a step in the right direction towards a more sustainable and environmentally friendly future.

12 Conclusion

The consumption of eco-friendly dyes continues to rise day by day as the demand for fresh and hygienic textiles rises. Currently, there is encouragement for the use of numerous natural pigments sourced from plants, animals and minerals in clothing, cosmetics, solar cell manufacturing and other applications. Several methods have been developed to extract the majority of colour-enhancing dyes from different sources. Research and development is working hard to keep up, creating ever more effective and secure solutions. Natural materials, such as those derived from minerals and animals, have drawn more attention as possible sources. Being one of the most dynamic fields, it needs to be updated with cutting-edge technologies. According to published research, the fabric's barrier qualities shield the skin from anti-ageing, minimise wrinkles and cool the skin without causing allergic reactions. This request is to researchers and scientists because there are so many commercially available natural sources on the market. But there is hardly any natural product that could satisfy all the requirements of consumers. So this is the main direction of new product development.

References

1. Deveoglu O, Sahinbaskan BY, Torgan E, Karadag R (2012) Investigation on colour, fastness properties and hplc-dad analysis of silk fibres dyed with rubiatinctorium l. and quercusithaburensisdecaisne. Colouration Technol 128(5):364–370
2. Devi AA (2014) Extraction of natural dyes from fungus—an alternate for textile dyeing. J Nat Sci Res 4(7):1–6
3. Sigurdson GT, Tang P, Giusti MM (2017) Natural colorants: food colorants from Natural sources. Annu Rev Food Sci Technol 8:261–280
4. Hussein SA, Barakat HH, Merfort I, Nawwar MA (1997) Tannins from the leaves of *Punica granatum*. Phytochemistry 45(4):819–823
5. Gershon H, Shanks L (1975) Fungitoxicity of 1, 4-naphthoquinones to *Candida albicans* and Trichophyton mentagrophytes. Can J Microbiol 21(9):1317–1321
6. Samanta AK, Agarwal P (2009) Application of Natural dyes on textiles. Indian J Fibre Text Res 34:384–399
7. Chandrashekar R, Rao SN (2012) Phytochemicalanalysis of ethanolic extract of leaves of *Clerodendrum viscosum* (EELCV). World J Pharm Pharm Sci (WJPPS) 1(3):1092–1099
8. Zachariah SM, Kumar NM, Darsana K, Gopal D, Thomas N, Ramkumar M, George N (2014) Phytochemical screening, formulation and evaluation of dried galls of Quercus Infectoria Oliv. Int J Pharm Sci Rev Res 26(1)

9. Chairman K, Jayamala M, Christy VR, Singh R (2015) Phytochemical screening and antimicrobial activity of *Curcuma longa* natural dye. General Med 3(2):171–182
10. Tiwari A, Srivastava M (2018) Phytochemicalanalysis of Methanolic dye extracts of some medicinal plants. Int J Appl Home Sci 5(9):939–942
11. Jothi D (2008) Extraction of Natural dyes from African marigold flower (*Tagetes erecta* L) for textile coloration. Autex Res J 8(2):49–53
12. Agarwal K (2009) Application of Natural dyes on textiles. Indian J Fibre Text Res 34:384–399
13. Singh BS, Panda AK, Jena J, Das T, Das NB (2012) Exploring sustainable technique on natural dye extraction from native plants for textile: identification of colourants, colourimetric analysis of dyed yarns and their antimicrobial evaluation. J Clean Prod 37:257–264
14. Ganapaty S, Vidyadhar K, Ganga-Rao B (2005) Antimicrobial activity of two Vitex species. Indian J Nat Prod 21:46–49
15. Kamboj A, Jose S, Singh A (2022) Antimicrobial activity of Natural dyes–a comprehensive review. J Nat Fibers 19(13):5380–5394
16. Kadolph S (2008) Natural dyes: a traditional craft experiencing new attention. Delta Kappa Gamma Bull 75(1):14
17. Chitra DT, Ravikumar R, Kavita N, SudarshanaDeepa V (2013) Impact of agitation for the colour removal from dye effluent using isolated fungal species. J Environ Res Develop 7(4A):1559–2156
18. Babel S, Gupta R, Mehta M (2016) Formulation of eco friendly print paste for block printing by Natural gum extracted from mango kernel and Cassia Tora seed with indigo dye. J Chem Biol Phys Sci 3:576–583
19. Rehman F-U, Adeel S, Qaiser S, Bhatti IA, Shahid M, Zuber M (2012) Dyeing behaviour of gamma irradiated cotton fabric using Lawson dye extracted from henna leaves (*Lawsonia inermis*). Radiat Phys Chem 81(11):1752–1756
20. Shanker R, Vankar PS (2007) Dyeing cotton, wool and silk with Hibiscus mutabilis (Gulzuba). Dyes Pigm 74(2):464–469
21. Hou X, Chen X, Cheng Y, Xu H, Chen L, Yang Y (2013) Dyeing and UV-protection properties of water extracts from orange peel. J Clean Prod 52:410–419. https://doi.org/10.1016/j.jclepro.2013.03.004
22. Montazer M, Parvinzadeh M (2007) Dyeing of wool with marigold and its properties. Fibers Polym 8:181–185
23. Krizova H (2015) Natural dyes. In: Kryštůfek M, Vik W (eds) Textile dyeing—theory and applications, 1st ed. TUL: Vysokoškolskýpodnik Liberec s.r.o., Studentská 2. Liberec, pp 317–334
24. Qicheng Z, Leu W, Sunghee K, Sunhua J, Menlong C (2003) Bio-dyes for wool. Textile Asia, pp 46–48
25. Tuli HS, Chaudhary P, Beniwal V, Sharma AK (2015) Microbial pigments as natural colour sources: current trends and future perspectives. Int J Food Sci Technol 52:4669–4678. https://doi.org/10.1007/s13197-014-1601-6
26. Heer K, Sharma S (2017) Microbial pigments as a natural colour: a review. Int J Pharm Sci Res 8:1913–1922. https://doi.org/10.13040/IJPSR.0975-8232.8(5).1913-22
27. Hernández VA, Galleguillos F, Thibaut R, Müller A (2019) Fungal dyes for textile applications: testing of industrial conditions for wool fabrics dyeing. J Text Inst 110(1):61–66
28. Mapari SA, Meyer AS, Thrane U, Frisvad JC (2009) Identification of potentially safe promising fungal cell factories for the production of polyketide Natural food colourants using chemotaxonomic rationale. Microbial Cell Factory 8:24
29. Rehman FU, Adeel S, Rafi S, Zia KM, Zuber M (2019) Environmentally friendly plant-based Natural dyes: extraction methodology and applications. Plant and human health, Volume 2: Phytochemistry and molecular aspects, pp 383–415
30. Rather LJ, Akhter S, Hassan QP (2018) Bioremediation: green and sustainable technology for textile effluent treatment. In: Sustainable innovations in textile chemistry and dyes, pp 75–91
31. Korankye O (2010) Extraction and application of plant dyes to serve as colourants for food and textiles. Doctoral dissertation

32. IUPAC (1997) The "Gold Book", 2nd ed. IUPAC, Compendium of chemical Terminology
33. Manhita A, Ferreira V, Vargas H, Ribeiro I, Candeias A, Teixeira D, Dias CB et al (2011). Enlightening the influence of mordant, dyeing technique and photodegradation on the colour hue of textiles dyed with madder–a chromatographic and spectrometric approach. Microchem J 98(1):82–90
34. Senthil Kumar P, Gunasundari E (2018) Sustainable wet processing—an alternative source for detoxifying supply chain in textiles. In: Detox fashion: sustainable chemistry and wet processing, pp 37–60
35. Feng Y, Liu Y, Xue L, Sun H, Guo Z, Zhang Y, Yang L (2017) Carboxylic acid functionalized sesame straw: a sustainable cost-effective bioadsorbent with superior dye adsorption capacity. Biores Technol 238:675–683
36. Alam MM, Rahman ML, Haque MZ (2007) Extraction of henna leaf dye and its dyeing effects on textile fibre. Bangla J Scien Indus Res 42(2):217–222
37. Baliarsingh S, Jena J, Das T, Das NB (2013) Role of cationic and anionic surfactants in textile dyeing with Natural dyes extracted from waste plant materials and their potential antimicrobial properties. Ind Crops Prod 50:618–624
38. Shah NC (1997) Traditional use of turmeric (*Curcuma longa*) in India. J Medi Aromatic Plant Sci 1997(19):954
39. Gulrajani ML, Deepti G (2011) Emerging techniques for functional finishing of textiles. Indian J Fibre Text Res 36:388
40. Prabavathy N, Shalini S, Balasundaraprabhu R, Velauthapillai D, Prasanna S, Muthukumarasamy N (2017) Enhancement in the photostability of Natural dyes for dye-sensitized solar cell (DSSC) applications: a review. Int J Energy Res 41(10):1372–1396
41. Kato H, Hata T, Tsukada M (2004) Potentialities of natural dyestuffs as antifeedants against varied carpet beetle, *Anthrenus verbasci*. Japan Agric Res Q 38(4):241–251
42. Arputharaj A, Raja ASM, Saxena S (2016) Developments in sustainable chemical processing of textiles. In: Green fashion, pp 217–252
43. Carvalho C, Santos G (2016) Sustainability and biotechnology—natural or bio dyes resources in textiles. J Textile Sci Eng 6(1):1–5
44. Bhuiyan MR, Islam A, Ali A, Islam MN (2017) Color and chemical constitution of Natural dye henna (*Lawsonia inermis* L.) and its application in the coloration of textiles. J Clean Prod 167:14–22
45. Chengaiah B, Rao MK, Kumar MK, Alagusundaram M, Chetty MC (2010) Medical importance of natural dyes—a review. Int J Pharm Tech Res 2(1):144–154
46. Gao Y, Cranston R (2008) Recentadvances in antimicrobial treatments of textiles. Text Res J 78(1):60–72
47. Kant R (2011) Textile dyeing industry an environmental hazard
48. di Poratelli F (2021) Fabric dyeing processes and their impact on the environment. Cikis Studio—Consulenza di sostenibilitÃ nella moda. https://www.cikis.studio/en/article/fabric-dyeing-processes-and-their-impact-on-the-environment
49. Kulczycki P (2021) The environmental impact of dyes in fashion. SANVT. https://us.sanvt.com/blogs/journal/the-environmental-impact-of-dyes-in-fashion

Optimized Resource Consumption

Anthima Ram and Anil Kumar

Abstract Garments don't come with an expiry date on their wash care label nor does the brand provide enough information on the garment with the hanging tags, garment labels, or the brand tag. Social media influencers or the trend change gives the confirmed indication that a particular style is now outdated and can't be worn again which now will be a part of landfill. The surge in fashion waste happens due to mindless purchases mindless consumption and mindless disposal. The accumulation of unused/unworn garments in the wardrobe becomes difficult to manage and adds up to a daily struggle of identifying a decent pair of garments to put on and feel satisfied in the second skin. Fast Fashion has been the time contributor in the short-lived fashion which directly ends in landfill causing climate change and increasing the carbon footprint. Capsule collections and minimalistic fashion lifestyle can play a key role in minimizing and optimizing the use of resources in fashion, making life easier with classic, and staple pieces of garments and ultimately reducing the waste dumpage in landfills.

Keywords Capsule wardrobe · Minimalism · Fast fashion · Sustainability · Consumption

1 Introduction

Today polyester through fast fashion is the most widely used textile, it exceeds cotton in production volume and now polyester usage in the global production of textiles is 57.7%. Polyester along with other synthetic materials, is made from crude petroleum oil which is an issue [1]. The extraction of oil from the environment and burning

A. Ram (✉)
Department of Fashion Design, School of Arts and Design, Woxsen University, Hyderabad 502345, India
e-mail: anthima.ram@woxsen.edu.in

A. Kumar
Panchkula, Haryana 134107, India

Sadhna et al. (eds.), *Climate Action Through Eco-Friendly Textiles*, SDGs and Textiles,
https://doi.org/10.1007/978-981-99-9856-2_8

of it leads to various climate change issues like rising sea levels, oceans becoming warmer, hotter temperatures, more health risks, and shifting weather.

The supply chain of virgin plastic as material is very opaque, the details about the extraction process of raw material till yarn production are deliberately kept discreet from the consumers. The dispersed dyes used to dye polyester textiles are based on the same complex molecular structure of the polyester fabric. The toxic wastewater from the textile factories gets flushed straight to the local water bodies and water supplies polluting the environment. The pollutant never degrades or dissolves in the water, which causes serious health hazards like cancer and respiratory disease.

When French couturiers like Chanel, Dior, Balmain, Givenchy, Calvin Klein, and Oscar De La Renta created a fashion collection using polyester, and polyester blended textiles and fabrics in the 1960s, it changed the perception of polyester. Being reliable and affordable at the same time [2], polyester became an integral part of the mainstream textile and fashion industry.

Replacing the entire drape, feel, and texture of actual silk, wool, and linen by mimicking their properties the innovative polyester got introduced into daily wear and household textiles as durable, lightweight, easy to wash, iron, and wear [3].

The revolution of wardrobes happened with the overvaluation of man-made fabrics over natural fabrics. Aesthetics of Modernists took over the domains of fashion, architecture, design, and products. This led to the arrival of bright colors, permanent pleating in fabrics, and more body-defining silhouettes in fashion.

With the rising middle class, the leisure and the mindset of buy, wear, and dispose (throw out) began with the globalization and evolution of fast fashion which eventually led to problems of environmental waste generation and pollution, exploitation of human labor, and depletion of natural resources [4].

As the system of fast fashion is works on polyester and polyester blend fabrics the clothes linger longer on the planet and hardly recyclable due to blends.

2 Garment Lifecycle: From Fiber to Landfill

The Garment Lifecycle starts with the cultivation of fiber-producing sources like cotton, silk, and sheep farming. From the fiber to fabric stage, it's crucial to be categorized in organically grown fiber, or chemically enhanced, with fertilizers and pesticides, synthetically designed for specific fashion [5].

The raw material for the fiber might be extracted from a single fiber or from a combination of several different fibers. The fiber is further divided into natural (animal and plant-based) and synthetic (manufactured) fibers, as well as inorganic, recycled, and regenerated cellulosic fibers [6]. Three techniques are used in the conversion of fiber into fabric: weaving for woven fabrics, knitting (flat and circular), and non-woven, which includes felting and binding [7]. Manufacturing phases for woven and knitted fabrics can be further broken down into two sub-stages: yarn production and fabric production. Spinning fibers into yarns, which are then turned into fabrics by knitting and weaving, is known as yarn production. Scouring is a step in the process of spinning fibers that removes contaminants while maintaining the quality of the yarn [8].

By this, the waste generation has started!

3 Role of Dyeing

Dyeing takes place in the fiber, yarn, and fabric stage. The wet processing stage, which comprises pre-treatment, dyeing, printing, and finishing of the cloth, is the last step in the manufacturing process. Indigo, black, red, and brown are among the colors that use a lot of water. To begin with, 3–12 mg of indigo dye is needed to make one pair of blue jeans. On average, 1 kg of Indigofera Tinctoria can extract 5 gm of blue indigo dye.

The garment or clothing manufacturing stage continues after the fabric is available and the quality condition of the fabric should match the requirement of the production house.

The processes and approaches used in the production of wearables. Following value-adding processes like printing, decoration, ironing, and packaging, involves pattern making, fabric cutting according to the pattern, sewing, and assembly [9]. Rarely, do dyeing operations take place at the stage of the garment as stated.

Retailing of the final ready garments is done through two ways conventional retailing and e-commerce retailing. Conventional retailing has a higher environmental impact than e-commerce retailing [10].

After a consumer purchases a garment, the lifecycle of the garment purely depends on the usage and maintenance of the garment. Washing, drying, ironing, and dry cleaning [11]. The number of washes decides the appeal for wearability based on factors such as fastness and retention of the color in 20 washes and the durability of the fabric and construction in 45 domestic washes.

Due to the unsustainable disposal practices used for post-consumer textiles and clothes, the end of the garment's useful life has a substantial negative influence on the environment. Utilization, recycling, giving, and incineration. Most of the used clothing that is donated from developed nations like the United States, the United Kingdom, and Australia is sent to underdeveloped nations like Kenya, Ghana, Rwanda, and other countries in southern Africa. Of these donated clothes, 80% are of poor quality or are not fit for re-wear and end up in landfills or are burned to ashes. Synthetic fibers can stay in the soil for hundreds of years without decaying, while natural and biobased fibers can release greenhouse gases when they are disposed of in landfills [12]. Figure 1 gives an overview of the garment lifecycle.

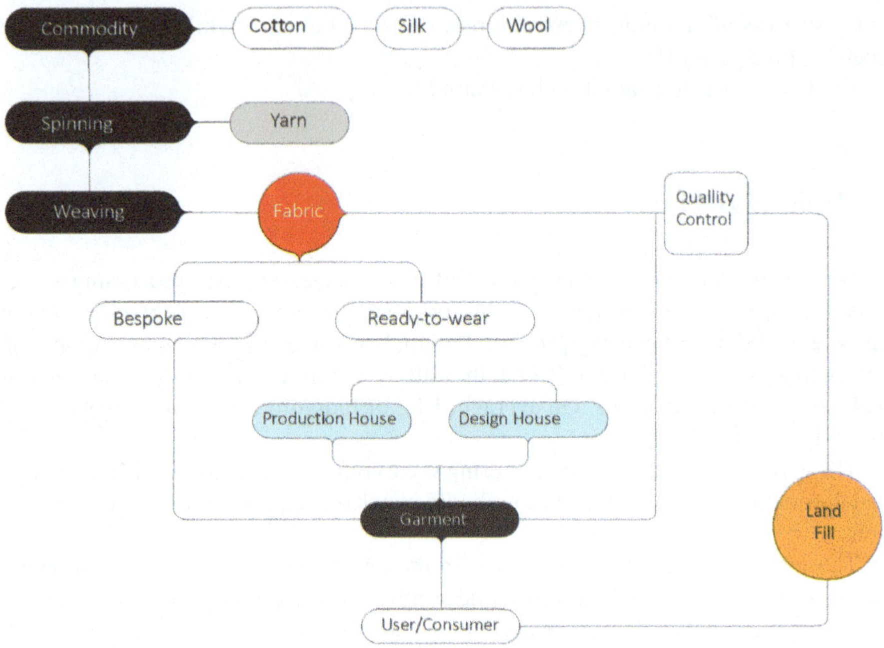

Fig. 1 The life cycle of a garment

4 Wardrobe! Manage My Fashion Ensemble

Our wardrobe is like our wallet. Why wallet?

The wallet is a place to keep things that are highly valued: literally, it contains money, credit/debit cards, a bus or train pass, identification IDs, and sentimental values. It symbolizes a conversation of life, health, and wealth. A wallet keeps everything flat and organized, working with the line and tailoring of the pants/trousers creating a smarter look and aesthetics. If we observe with close attention the design structure of the wallet plays a prime role in storing all valuables in different compartments, easy access to keep and retrieve things safely in any situation, and lastly mindfully keeping the necessary valuables only to carry, use, and interact.

In a similar line of thought, the wardrobe and the standing closet enable a person to store outfits, accessories, and valuables as per usage and keep garments and products safe in various compartments/partitions. A well-organized wardrobe helps in finding and picking up the right outfit in half the time you need there and then. It maximizes and helps in managing the wardrobe space, hanging rails with sorted space with specific fashion products and drawers to accommodate all other products of your closet/wardrobe.

Everything has a place, and you can quickly locate the things you need rather than wasting time frantically looking for them, which saves time. Mindful planning and customized layout knowing what it is that you want to store in there and how much space you have in your wardrobe gives you, ideas to place your items and utilize the most of that wardrobe space.

We all know the symbolic meaning, purpose, function, and use of a wallet. But do we ever think about our wardrobe in terms of stacked up clothes, garments, bags, shoes, and jewelry from which we hardly utilize the 20% in our daily life and forget what happens to the rest of the 80%?

Closet or even the word wardrobe was not discussed much about two decades before as now we see it as part of the mainstream issue directly relating to fashion landfill dump-age. Organizing of wardrobe is often looked at as a frustrating and boring activity. Though in the busy schedule of life, a few people sometimes avail the paid service to manage their wardrobes from a professional wardrobe manager.

Consumption of fashion, especially fast fashion has turned to owning garments that are bought under the influence of being trendy and to be seen fashion updated without because how many times it will be worn and then forgotten under the piles of unwanted clothing in the wardrobe which were once trendy in one season and treasured upon. Rapid technological advancement has doubled clothing production and garment usage lifetime has decreased.

5 Birth of New Era, New Professions, New Fashion

The birth of fast fashion happened due to the emergence of fossil fuels, the Industrial Revolution, and capitalism. As production from textiles ramped up over the nineteenth and twentieth centuries—both from mechanization and the exploitation of the working-class capitalists required people to buy all the products they were producing. The internet, technical advancement, and globalization have accelerated the pace of fashion's updating, which has given rise to new markets and industries that are significantly distinct from the traditional fashion sector. The latter appears to have a shorter fashion cycle because it has been accepted and grown based on demand, consumed based on occasion, displayed, and discarded more quickly than the traditional fashion life cycle upon trend fading [13].

The dictatorial fashion trends were the by-product of this revolution; the wielding of fashion's cyclical nature became much more acute as industrial capitalism took hold of the global fashion market. Formerly fashion shows, catalogs, and advertisements display what a consumer could be if they own just that pair of pants from Zara shoes from Nike, or a t-shirt from H&M. In a capitalist system where endless hours of meaningless or destructive work grind people into unhappiness, and these same people who become potentially unhappy mass consumers seek consumption as an easy to achieve happiness.

The introduction, acceptance, peak, and decline of a particular trend or style are all parts of the fashion life cycle. The cycle begins with innovation and intervention, which together make up the initial stage, known as the foundation stage. After the successful adoption of a trend, the fashion reaches its zenith in the acceptance stage. Following that, as time passes and fewer individuals adopt the fashion, it progressively loses favor and eventually fades from public view.

To meet consumer demand, an organization's strategy known as "fast fashion" reduces the lead time duration for launching new trendy products into outlets and the procedures involved in the buying cycle. When it is at its peak by being CHEAP, CATER, and DOMINANCE [14].

CHEAP: A rapidly evolving, low-cost fashion style defines our era. For the fashion industry to more readily delocalize everything from manufacturing to distribution and become more globally developed than other industries, this is necessary. Fast fashion's main goal is to democratize fashion through its affordable prices and quick turnaround times, hence inexpensive plays a big part in this sector.

CATER: Brands like Zara and H&M implement strategies to keep their product shelf-life shorter so customers keep visiting stores every week to not miss out on new fashion products.

This strategy works based on the "response to the consumer preference." Based on the response if a product doesn't sell well then, it's removed from the store and replaced with a new design. Zara decided to cater to the public by quickly rejecting their design. Zara lovers are again encouraged to return for additional shopping because no design is kept in stock for longer than four weeks.

DOMINANCE: To design and manufacture exclusive hit products that appeal to the newest consumer trends and leverage short production lead times to supply uncertain demands, fast fashion has coupled quick response production capability with better product design proficiency. Along with addressing customer preferences, this is done [15].

To tap into the emotional needs of the consumer fast fashion brands study and analyze their consumer's needs, wants, desires, and aspirations based on the consumer's understanding of their fashion sense, style, and self-image. These attributes are further broken down into customer shopping behavior and purchasing frequency, source of fashion trend information, influence, awareness, and preference, before-purchase need generation, selection and purchase decision-making, after-purchase utilization duration, and divestment of product.

Fast fashion brands create needs, desires, and social validation through social media platforms and encourage their purchasing behavior in various ways.

Created needs and desire to own a new piece of trendy garment for every occasion, party, and event in life has always kept the consumer thinking, visualizing, and manifesting to look stylish, fashion-forward, upbeat with the latest trend, rising on the ladder of status and ultimately look like a celebrity or model [16]. Major strategies commonly employed by the brands are:

5.1. **Trend-driven marketing**: Brands emphasize the latest trends and create a sense of urgency to stay in style. They utilize social media, influencer marketing, and posting their popular customers' good reviews and stories on the e-commerce websites making consumers feel the need to constantly update their wardrobe.

5.2. **Instant gratification**: Fast fashion thrives on providing immediate satisfaction. Allowing the consumer to experience the thrill of acquiring new clothing quickly and easily.
5.3. **Low prices and affordability**: Creating a perception of affordability that appeals to the consumers who seek budget-friendly options and fosters the idea that they can acquire more for less.
5.4. **Social validation**: Featuring ideal and attractive models, aspirational imagery, and stories of confident individuals, they aim to make consumers feel that purchasing and owning their product will enhance their image and social acceptance.
5.5. **Emotional connection to the brand**: Through narratives of consumer's values, aspirations, and self-expression, making them feel a sense of belonging and identification with the brand.
5.6. **Convenience and accessibility**: Prioritizing convenience, offering online shopping, fast shipping, and hassle-free returns. By removing barriers to purchase, and satisfactory shopping with ease and effortless experience.

The strategies keep improvising with better AI-based recommendations from past purchase processes, decision-making with chatbots, and virtual trial rooms are effective in driving sales. Once the garment is worn for hardly three weeks, the trend dies off faster than acquisition leading to a surge in waste as piles of garments unused in the wardrobe eventually end up straight in the landfills, or in charity houses/thrift stores where only a handful of garment pieces are worthy enough to be re-worn with proper washing, repair, and invisible mending.

6 52 Seasons, Luxury Conglomerates Doubling Down to Sustainability and Emergence of Sustainable Fashion Capitals

Major fast fashion brands design collections following the 52 season all over the year, which means new trendy collections getting stocked up in stores around the world every week. In the United States alone a customer buys approx. 68 pieces of clothing in a year and on average wears a garment just for 7 times before discarding it. The created need by the brands to look fashionable, and trendy and be aware and part of the latest styles becomes one the reason to buy fresh new garments.

Four seasons are recognized as the official fashion seasons. However, there were only two fashion seasons back then: spring/summer and fall/winter. Fast-forwarding from the past to the present, certain fashion houses and companies create 52 microseasons annually [17]. Every week, new trends appear, and clients' goals are predetermined to buy as much clothing as quickly as they can.

In the previous era of two seasons, the fashion was warmer clothing in colder weather, poppier and brighter colors in the sun, and more functional designs for sports or inclement weather.

In contrast to the Autumn Winter, which lasts from July to December, the Spring Summer starts in January and lasts until June. Pre-fall collections are offered before the Autumn/Winter collections come, and resort collections are accessible from October through December.

The idea of the season ends up being much more sophisticated because it serves the function and needs to be related to the weather and climate. The pace and duration of designing, developing, producing, marketing, launching, and selling the newest collections are determined by the seasons, which act as a global metronome for the fashion industry.

Other companies, such as Pantone and Worth Global Style Network, play a crucial role in determining the colors used and trend forecast predictions from designer runway studies used in practically everything we see around us. These agencies, in addition to the major ones in the publishing, advertising, logistics, and distribution sectors.

Despite being driven by innovation, the fashion industry nevertheless adheres to traditional commercial methods, therefore designers started defying the rules and exhibiting their range between the two recognized seasons. These were referred to as "Pre-collections" and frequently served as a designer's ready-to-wear collection before the introduction of significant exclusive products. Since they remain in stores longer, are frequently less stunning, and offer more wearable things for consumers to choose from, fashion manufacturers typically make more money from their pre-collection. Customers can purchase pre-collections from May to July in advance of the September Spring Summer fashion season and once more in November or December in advance of the next year's Autumn Winter range.

The major seasons displayed during Fashion Week to the press in September are the spring and summer collections. In contrast, the winter of 2024 will see the debut of the autumn/winter collections in February. Fashion brands and labels frequently shoot resort and pre-fall collections exclusively for the press, fashion editors who choose samples and photograph them for editorials, and retail customers who can view the collection and choose which items to buy. These collections do not have a dedicated runway showcase.

The advancement in synthetic fabric technology and rapid manufacturing methods revolutionized the industry.

Fast fashion labels introduced more styles and trendy clothing and accessories at affordable prices which also contributed to the evolution of 52 micro seasons. With the fashion brands making a fundamental shift in their profit-making approach the quality is compromised by charging high margins for a selected number of high-quality items and would add a slight margin for an unlimited amount of cheaply and quickly produced collection garments. Weekly, fashion stores like Zara, and H&M change their appearance with complete inventory overhauls and with slick advertising aimed to stroke the hunger of buyers around the cloak [18].

An unlimited supply of new trendy styles every week with dictatorial advertising has led to rapid mindless consumption and acquisition causing undervalue and reducing their appreciation for purchases. Customers are conditioned to behave that they should always be trendy, and fashion updated. They need to stay on the "hamster wheel" of weekly purchasing to maintain their level of happiness. In addition to the psychological effects are the socio-economic ones like the source of happiness and satisfaction, climbing the ladder of social class by imitating upper-class lifestyle leading to a slow death of popular fashion trends.

Consumer activists' criticism of labels for their scarcity of supply-chain visibility, together with claims of workers and cruelty towards animals, is resulting in growing tensions in the areas of alarming climate change, fast fashion consumption and disposal, and luxury organizations. Dior, Chanel, and Dolce & Gabbana are being criticized for their supply chains and their unlawful and ineffectual policies, while fashion tycoons like Prada, Hermes, Gucci, and Louis Vitton are accused of unethical and illegal actions toward animals to make handbags and footwears from their exotic-skins. Organizations are being forced to exercise caution as a result of a combination of growing awareness of socially unethical and unlawful corporate operations and intense public pressure.

The first complete social and environmental report from LVMH, which also included a new sustainability strategy called Life 360, was published in May 2021. Through a combination of communication with Maisons, employees, and students as well as rigorous effect measures, the company's Life 360 program, with specific targets established for 2023, 2026, and 2023, strives to create sustainable products that positively enhance biodiversity and battle climate change.

The company's Life 360 program with specific targets set for 2023, 2026, and 2023 aims to create sustainable products that positively impact biodiversity and combat climate change through a combination of dialogue with Maisons, employees, and students, as well as precise impact measurements [19].

The Life 360 program focuses on biodiversity, circular design, traceability/transparency, and climate change in addition to cutting-edge research projects such as agroforestry programs, keratin-based fiber development projects, recycling, and upcycling. Last but not least, LVMH supports cultural diversity and artistic innovation as part of its commitment to social responsibility. A global Diversity and inclusion strategy has been introduced by the organization to achieve gender parity and equitable pay. increasing the share of disabled individuals in the global workforce to 2%.

On the other side, Kering unveiled its 2025 sustainability strategy in its roadmap for the years 2017 to 2025, intending to cut its environmental footprint by 40%. To promote luxury with longevity, the company takes a holistic approach through sustainable innovation, focusing on circularity, new materials, innovation laboratories, production techniques, new business models, as well as upcycling, recycling, and regeneration. The three pillars of the group's strategy are CREATE, CARE, and COLLABORATE. With the use of cutting-edge tools, it seeks to lessen its environmental impact (Care), encourages diversity strives to become an employer of choice (Collaborate), and provides original solutions utilizing an open-source methodology (Create) [19]. In January, Kering joined businesses like Adidas, PVH Corp., a major US retailer, and Arvind Limited, an Indian textile manufacturer, in partnering with

Fashion for Good. Through a program known as the D(R)YE Factory of Future, the partnership is anticipated to lessen the environmental impact of material processing.

Kering partnered with the London College of Fashion on the Massive Open Online Course to incorporate sustainability as a central theme into fashion education and promote the wide implementation of more eco-friendly practices. The course, "Sustainability & Fashion: Comprehending Luxury Fashion in a Changing World," was designed by sustainability experts from Kering and LCF Educational institutions to address the theoretical and real-world business difficulties that luxury fashion companies face when implementing environmentally friendly policies [20].

This shows that sustainability and luxury may not be compatible because ostentation and excessive consumption have historically been associated with luxury. Sustainable spending is linked to restraint and ethics, whereas luxury values are linked to personal delight. However, there are several possible intersections between luxury and sustainability. Selective retailing channels, the creation of limited editions, and markets where luxury goods are viewed as both uncommon and enduring highlight the scarcity of premium brands frequently. As a result, customers perceive them as being more socially concerned. Additionally, luxury brands advertise their products as timeless, suggesting a link between luxury and sustainability since long-lasting grandeur is unaffected by passing trends.

Fashion capital "PARIS" the city in 2019 announced a five-year plan aiming to be the sustainable fashion capital of the world [21].

The initiative is supported by both The Galleries Lafayette and LVMH Moet Hennessy Louis Vuitton. Participants include the Ellen McArthur FOUNDATION, the IFM, Eyes on Talent, the Federation de la Haute Couture et de la Mode, and Vogue incubator Les Ateliers de Paris. To give out a roadmap of the measures that may be taken to be more environmentally responsible, experts from the fashion industry, businesses, entrepreneurs, designers, and specialists have founded the open community known as Paris Good Fashion. The three primary areas on which committee members concentrate their efforts are the development of a circular economy, enhancing sourcing and traceability, and aiming to make some activities, such as distribution, energy use, and communication, more sustainable.

New fashion capitals are emerging in European countries with a mission to make more conscious efforts and to be more responsible toward changing the linear model of fashion product/business, process, waste generation, consumption, and disposal which contributes to climate change which is harmful to the people, planet, and profit.

Berlin and Copenhagen are the new capitals of sustainable fashion, cool, and street styles. These cities are coming under the spotlight and innovation to set new standards to address the challenges of catwalk presentation, which creates a great amount of waste from paper invites to water bottles and to the single-use materials going into gigantic show sets accounting for about **241,000 tons** of carbon emission in a year.

The Berlin Fashion Week also set a multi-faceted Berlin fashion Summit to promote Activism, Innovation, and Experimentation. With an agenda to focus on environmental consciousness, fair pay, slow fashion, and responsible practices, Regenerative Business and Regenerative Culture called for economic, cultural, and ecological actions. Berlin sustainable brands presented themselves through immersive collection experiences at the Berlin Metaverse—an experimental platform showing the newest possibilities within fashion tech and how it aligns with circularity, exploration of gender fluidity, and diversity in their collections.

Copenhagen Fashion Week introduced a Sustainability Action Plan with 18 minimum standards that cover six key areas, from design to show production and working conditions, set targets to reduce its environmental impacts by cutting 50% of its emissions, which will be reviewed on an annual basis to reflect the changes happening in the industry to reduce industry's impact on the planet. Nordic Future fashion brands and their sustainability goals align with CPHFW and promote them globally [22].

7 Role of the IT Revolution in Surge of Waste Volume

In the digital world of fashion, apps have made shopping hassle-free, experiential, and even stress-relieving for some consumers.

Indian e-commerce apps like Myntra, Ajio, Flipkart, and Amazon offer and deliver similar shopping experiences along with easy return policies creating an opportunity for unlimited access to browse, put items of interest in a wish list/cart, and buy it now or later on preference. Fashion brand applications overly simplified and engaging user interfaces allow consumer to browse, filter and select, purchase, payment, get to choose delivery place and hours, and return hours altogether in real quick time and steps which keeps the consumer hooked on shopping mindlessly by providing temporary happiness and satisfaction.

Notification of new season collection launches, sales/discount offers, recommendations based on similar products, and cheap price points are all aimed by these apps like Zara, Shein, Uniqlo, and H&M with added features along with filters, customer product reviews which almost try to give satisfactory shopping experiences to the consumer trick them into buying more clothes every month.

Fashion houses and labels are now utilizing technology (AI) to strengthen client purchase experiences, analyze preferences based on data, expand sales, predict and recommend trends, and guidance on inventory issues. Virtual trial/fitting rooms are the next step in the evolution of retail experiences that merge the physical and digital worlds. Customers' decision-making will improve and the store's profits will rise if there is more choice and clarity during the purchasing process. Chatbots and touchscreens are being used in malls and stores to improve customer service and provide individualized product recommendations. The employment of artificial intelligence chat GPT technology to enhance the user experience is now practically a given while visiting the website of a fashion company [23].

An automated method for designing one's wardrobe has been developed by the British clothing line "STITCH FIX" which uses analytics to track customer purchases and present them with a virtual wardrobe. The platform also makes it easier for customers to put together outfits from their closets and even make selections from more than 15,000 stores. With the help of an online fit engine, the customization platform TRUEFIT helps customers find a good fit with well-known brands and cutting-edge fashion trends.

Live data analytics software is offered by platforms like Edited and Intelligence Node so that its retailer customers can instantly access all market data and synthesize the entire global market. Our daily lives now include live video streaming. From virtual events to health, Instagram commerce has dominated the post-COVID industry in 2022. Thanks to 5G, high-definition streaming video formats are now feasible.

The lack of supply to meet the rising global demand for garments during the COVID-19 outbreak caused the fashion industry to suffer. During the COVID pandemic lockdowns, shoppers drastically curtailed their fashion purchases and classified them as "unimportant" categories of consumer goods. Due to restricted access and limits on everyday necessities for lifestyle and welfare, long periods of complete and partial lockdowns led to the clearing out of closets and an increase in donations of clothing and textiles.

The rise of middle-class disposable income has significant implications for fast fashion brands. As it grows, there will be an increase in consumer spending as we have seen with Shein, Zara, and H&M, which could lead to important investment opportunities in brands that cater to this demographic. In addition, the growing numbers of middle-class youth will drive demand for a range of other products and services, including health care, education, and financial services [24].

Anything and everything has its pros and cons, nowadays as I/we have few negative viewpoints related to shopping, and enjoying fashion. May reduce the volume of waste and can direct the choices of human-related garments and fashion. In the coming decade, AI and AR can grow so much that our ecosystem of fashion and choices of fashion become so accurate with the help of suggestions. We only end up buying the right clothes, and garments, and picking and following the right trend as we are exponentially growing in this tech tomorrow, we might shift to a very different tangent to express our fashion digitally and a new self-expression may evolve say *Digital fashion appearance* [23].

Anyone (known person) may get relief from wearing an outfit to satisfy certain appearance parameters at public places like airports, beaches, temples, etc. According to current trends and practices (in times of filters) in my vision we might end up expressing our fashion digitally with the ease and comfort of bedtime clothing. In the future, AI may replace stylists partially and it will research and suggest what to wear can be the style guide, in addition to wardrobe manager.

So, generating an ensemble for an occasion or event that one can wear digitally because all that matters is pictures and videos. AI can generate an ensemble and the filter which can be layered onto cameras of media and paparazzi, and hence digital fashion comes into play as a new way to express yourself. As I have tapped into a new portal of imagination for fashion expression, readers can also float into their imagination and just can explore and relish the vivid spectrum of freedom for fashion. We can build a universe where the needs of every stakeholder of fashion enthusiasts can be satisfied, and even a new set of audiences can be seen. Fashion designers will gain more degree of freedom and fashion shows which we see today can be restructured with endless permutations and combinations.

8 Capsule Wardrobe and Minimalism, a Conscious Shift to Responsible Consumption

Minimalism originated in the United States in the 1960s as an art form centered around simple geometric shapes. The art form evolved into the minimalist lifestyle, where you aim to surround oneself only with things you need, instead of things they want.

The need for more functional fashions and the resulting fabric restrictions brought about by World War II necessitated a rise in the standardization of apparel production across the board. After the war, middle-class consumers were more accepting of the mass-produced apparel and styles as the accustomed homogeneity became the norm. To follow these new styles in the 1960s, young customers abandoned the sartorial customs of the pre-World War I clothing generations and embraced inexpensively created apparel [25]. Fashion companies discovered ways to meet the rising demand for reasonably priced, trendy clothing, which resulted in the opening of numerous textile mills and manufacturing facilities throughout developing nations. This allowed European and American companies to take advantage of the low-cost labor by outsourcing and save millions of dollars.

A key feature of 1960s culture was expendability, which was both a physical characteristic of all objects and a representation of faith and dependence in the modern era. Obsolescence the idea of having the awareness that the currently existing version of a given product will become outdated or out of fashion in the known period became accepted and often positively celebrated in the 1960s due to the post-war mass-production and standardization.

By offering a superior replacement model or purposefully constructing a product to stop performing a necessary function after a predetermined amount of time, style obsolescence is achieved, increasing consumer desire for newer products and driving out older ones. The Nylon stockings were replaced regularly due to the snagging, snaring, or laddering of the stockings.

Disposal of products only took place when there was a real need for a new trendy or desirable style for consumption. Fashion clothing or fashion products didn't come with a designed expiry date for use. The added value through aesthetics and attractive appearance termed as a trend-forward (eye appeal) makes the consumers eagerly wait for the next upcoming trend.

Fast fashion brands like H&M, Zara, Topshop, and Primark strategized on planned obsolescence to keep their consumer mindlessly hooked to purchase their products almost every week which favors the shooting upwards of the sales curve and generation of "progressive waste or creative waste" [19].

Overconsumption, unnecessary purchases, and a disordered wardrobe all contributed to the daily battle to locate the appropriate pair of clothes to wear for day-to-day life. The major population that faces this problem is the women. Despite having a plethora of clothes stored in their wardrobes/closets, ladies experience a significant deal of difficulty in deciding what to do during the day. Overwhelming options tend to confound and complicate the straightforward action of dressing up. After witnessing her customers wasting assets on too much clothing that was not sustainable in the long run owing to bad fabric quality, cheap construction, lack of comfort, fit, and timeless design, author Susie Faux opened her London's West End store called "Wardrobe" in the 1970s. Susie Faux's concept was to acquire traditional, high-quality pieces that would last for years without sacrificing fit, comfort, or quality [26].

In her book "**WARDROBE**" Susie Faux put down:

> **"If we apply capsule to everyone, we have to ask: What is her or his lifestyle? What is the most important piece of that lifestyle? That might be a perfectly tailored dress to see friends in; or a beautiful jacket for the office. For others, it might be nice jeans and tops. But for the average person, women and men who go to work need the best clothes they can."**

Further in her work toward creating fashion for women, she helped women look their best by integrating their personality and lifestyle into their style with confidence. Susie Faux invented the concept of "The capsule Wardrobe" which is a modest number of "key" or "Staple" pieces in complementing colors that can be combined and matched to create a look for any occasion without owning too many clothes. The collection's necessities included "2 sets of pants, a frock-like dress or skirt, a jacket, a coat, a knitted garment, two pairs of footwear, and 2 carrier bags."

Later, in 1985, Donna Karan the American fashion designer debuted her very first collection for her eponymous label. The designer introduced America to her "Seven Easy Pieces" line, an improvisation on the "Capsule Wardrobe" to fill "a void marketplace" for an attractive and functional wardrobe tailored for working women. Donna Karan introduced a wardrobe based on black, polished, and subtly provocative, made in luxurious jersey and cashmere sweater, layered upon a bodysuit with a tailored jacket, white shirt, skirts, pants, and tie suit [27].

In the 1990s, minimalism as a ready-to-wear philosophy dominated the fashion industry, appearing on the runways and reappearing in design trends now and then. Minimalistic fashion consumption has derived various lifestyle changes that aim to reduce the effect of climate change, and depletion of natural resources. Mass consumption and fast consumption have made customers do mindless buying and mindless disposal of cheap fashion. Low-cost fashion products made from petroleum byproducts have flooded the market.

Apple's iPhone design philosophy represents minimalistic design aesthetics that have changed the way humans consume, line, and behave following intensively simplified intelligence. With the most successful sales in history, the most valuable brand in history, and the most iconic successful manufacturer of minimalist product designs, consumers eagerly await the new iteration of the iPhone.

Minimalism at its core is for simplifying the brain's perception through basic simple shapes, and plain colors visually with high quality and completion without unnecessary attributes or elements. "Simplifying perception is more important than simplifying form," stated Minimalist artist Donald Judd's artwork. When a person sees something, he or she does not retain every feature as it is, but rather simplifies the shape and saves it in memory without putting much effort into recognizing processing, and storing information, as Gestalt Law suggests. Energy consumption optimization is a fundamental feature of the world of nature.

The current culture has access to an endless amount of information, and the processing of information has expanded exponentially, resulting in less energy and perception spent on each piece of information. Accessing complex information more intuitively and concisely is made easier and more intuitive by minimalist design [28].

Minimalist fashion manifests itself as a new aesthetic form, rather than as a desire to live with less. Minimalist clothes encourage the wearer to appreciate materiality, silhouettes, and wearability while still looking and feeling great. The minimal dressing is frequently comprised of an uncluttered color palette of black, white, greys, neutrals, earthy tones, and the occasional splash of maximal colors. Minimalist jewelry can be delicate, elegant, and subtle, or it can be big, bulky, and colorful, but it is always undeniably personal. The appearance of the modern contemporary minimalist is timeless and subtle, with an unfussy elegance that looks to be effortless. Investing in pieces of outstanding workmanship, finishing, and fit is essential to embracing minimalism style. Personal taste becomes apparent yet intellectual when items fit perfectly.

Minimalism and sustainability entail living with existing garments in one's wardrobe and acquiring them only when necessary. Voluntary simplicity is a reaction to excessive consumption during the "peak stuff" period. Consumption that is calculated and conscious, yet with a fashionable and stylish edge [29].

Because the emphasis is on garment quality rather than meaningless quantity, minimalist fashion consumption promotes circular design and circular economy concepts. Treating the root cause of excessive fast-fashion consumption by avoiding initiating the cycle of consumption, usage, disposal, and re-use in the first place. Journalists and bloggers have largely speculated about the growing popularity of minimalist lifestyles and their potential effects on sustainability; however, neither a clear empirical definition nor solid scientific proof of these effects has yet been definitively proven in the scientific literature [30].

9 Greenwashing and Ecolabels

Knowing the meaning of the new term "Greenwashing" when a fashion house expends more time and finance in portraying itself as eco-friendly than it does on literally minimizing its impact on the environment, this practice is known as "greenwashing." It's a dishonest marketing gimmick designed to trick consumers who choose to buy products and services from environmentally conscious businesses.

Jay Westerveld an Environmentalist from New York in 1986 coined the term "greenwashing" in an essay in which he criticized the Hospitality industry's practice of displaying signs in rooms urging towel reusing to "save the environment."

The Changing Markets Foundation found that 60% of the assertions made by the major fashion brands in America, Europe, and Britain might be categorized as "misleading" and "dubious."

To stop the quick depletion of natural resources and slow climate change, fashion brands are spending millions of dollars on campaigns, advertisements, and marketing strategies to persuade and entice consumers into thinking that their fashion products are made with fair trade and environmentally friendly practices [31].

The definition of sustainability stated by the fashion brands differs as per the brand image and philosophy.

Greenwashing is based on highlighting the minor good and positive parts of their practice while obscuring the rest that have a bad impact on the climate to showcase themselves as highly sustainable than they are. Claims of greenwashing concern the sustainability of synthetic fibers or the advertising of recycled plastic as a new material for clothing. Fashion firms utilize eco-labeling or sustainability certifications to win the trust of their target audience. Through eco-labeling or certification, a mark of supreme quality, superior value, thoughtful manufacture, and authenticity is meant to be changed or produced [32]. The capacity to influence customers' willingness to spend more for specific goods is what gives the Green Movement its reputation.

According to the Changing Markets Foundation's most recent report top 10 certifications, labels, and voluntary industry initiatives in the fashion sector include EU Ecolabels, MacArthur, Cradle to cradle, OEKO-TEX, SAC and Higg Index, BLUESIGN, Textile Exchange, The Microfiber Consortium, WRAP and ZDHC. These projects all share a lack of accountability, weakened independence, and lack of transparency, which is unfortunate because the majority of them are just greenwashing tactics for brands [33].

The use of bogus eco-labeling and certificates by businesses to position themselves as ethical, ecological, and eco-friendly has resulted in a growing number of issues that the fashion industry is currently grappling with. Regulators have uncovered major corporations like H&M and Decathlon claiming false "green" commitments. As an element of a group initiative to build its Green Declare Code, regulators in the United Kingdom have looked into comparable claims stated by Boohoo, ASOS, and George. When it comes to materials, recycled polyester in particular is evolving into a key environmental problem for the fashion sector. The clothes are being referred to as "recycled" even though there is no evidence that they are part of a circular system for apparel.

Ecolabels, this one new emerging practice in support of minimalism and reducing waste volume. All the fashion products must be evaluated for sustainability [34]. The sustainability movement has gained support from the government and every business. The government is promoting sustainability standards and ecolabels that encourage the ecologically safe and eco-friendly properties of textiles. Ecolabelling is one of the strategies that has been established to offer fragments to demonstrate and establish the green credentials of a garment that supports consumers in making educated decisions about their buying. The manufacturer must adhere to the ecolabel's requirements, according to the ecolabelling system, to be certified for producing a "Green" product [35].

There is a lack of knowledge regarding the lifecycle of fashion items and eco-friendly fashion labels. It should go without saying that the first step is to raise customer consciousness and receptivity to change by encouraging them to purchase products made from natural fibers without the use of harmful or toxic chemicals as well as clothing manufactured in nations with stricter environmental regulations for textile factories. Most well-intentioned efforts to reduce the environmental impact of the fashion business concentrate on people's individual lifestyles and moralistic personal attitudes to purchasing, caring for, and discarding clothing. However, a garment's ecological footprint is largely determined by its manufacturing (including transit), which is thought to be responsible for 80% of the impact. Even conservative estimates of this computation place the number above 60%. This indicates that decisions made about production in corporate supply chains are to blame for the great majority of suffering caused by the fashion industry.

10 Conclusion

By promoting awareness about sustainability standards and eco-labels, we can encourage individuals to make more informed choices when it comes to their fashion consumption. This increased awareness benefits not only the environment but also fosters a culture of consumption that is more responsible and conscious. Furthermore, the integration of AI-based technology for sorting wardrobes and reducing waste from post-consumer fashion consumption represents a significant step forward in our pursuit of sustainable fashion practices. AI can assist individuals in curating their wardrobes, ensuring that each piece is cherished and utilized to its fullest extent, thus minimizing the need for constant new purchases and reducing the fashion industry's ecological footprint. A crucial aspect of this sustainable fashion movement is the promotion of eco-textiles. These textiles offer a viable alternative to traditional fabrics, as they are produced with lower environmental impact, often using recycled materials or sustainable farming practices. By encouraging the use of eco-textiles, we can further reduce the fashion industry's negative environmental consequences.

The adoption of capsule wardrobes, minimal fashion, awareness of sustainability standards, AI technology, and the promotion of eco-textiles all play pivotal roles in the journey towards a more sustainable and responsible fashion industry. Our research underscores the importance of these initiatives and the potential they hold to minimize resource consumption, reduce waste, and promote a brighter and more eco-conscious future for the fashion world. We must continue to advocate for and implement these strategies to drive lasting change in the way we consume and appreciate fashion. Let's understand a future projection and what measures could to reduce waste volume and optimize the use of garments. After the IT revolution next breakthrough was in AI.

The further future scope of research should take advantage of Artificial Intelligence to propagate, promote, and recommend consumers to sort their fashion choices to reduce waste generation from their wardrobe itself which holds a huge potential to optimize the usage of already existing garments and stop the mindless purchase and mindless throwaway culture which is a direct result of fast fashion. We can plan an ecosystem that can assist people to be in trend and suggest they donate the garments to pursue fashion trends guilt-free. A direct donation layout can be designed to benefit the population in need. AI applications can guide fashion enthusiast to manage their wardrobe and garments and keep on suggesting whether to keep them, store them, discard them, or donate them. Imagine if we can cut off 20% of the volume which ends up in the land field after use.

References

1. Smelik A (2023) Polyester: a cultural history. Fashion Pract 15(2):279–299. https://doi.org/10.1080/17569370.2023.2196158
2. Bramel S, Fauque C (2001). The genius of the permanent fold: 100 years of textile modernity
3. Brunnschweiler D, Hearle J (1993) Polyester: tomorrow's ideas and profits, 50 years of achievement. Textile Institute
4. Meyer-Larsen W (1972) Chemiefasern. Rowohlt, Germany
5. Esteve-Turrillas FA, de la Guardia M (2017) Environmental impact of recover cotton in textile industry. Resour Conserv Recycl 116:107–115. https://doi.org/10.1016/j.resconrec.2016.09.034
6. Saheed H (2017) Prospects for the textile and clothing industry in India, 2017. Global Apparel Markets. https://textilesintelligence.com/tisgam/?pageid=3&repaid=TISGAM&issueid=34&artid=2232
7. Bevilacqua M, Ciarapica FE, Mazzuto G, Paciarotti C (2014) Environmental analysis of a cotton yarn supply chain. J Clean Prod 82:154–165. https://doi.org/10.1016/j.jclepro.2014.06.082
8. De Saxce M, Pesnel S, Perwuelz A (2012) LCA of bed sheets—some relevant parameters for lifetime assessment. J Clean Prod 37:221–228. https://doi.org/10.1016/j.jclepro.2012.07.012
9. Fung Y, Choi T (2017) Product development process of an international luxury fashion brand: implications to Hong Kong fashion trading and manufacturing companies. In: Springer series in fashion business, pp 27–41. https://doi.org/10.1007/978-981-10-70075-5_2
10. Wrigley N, Lambiri D, Astbury G, Dolega L, Hart C, Reeves C, Thurstain-Goodwin M, Wood S (2015) British high streets: from crisis to recovery? A comprehensive revview of the evidence. In: University of Southhampton ebooks. https://eprints.soton.ac.uk/375492/
11. Druckman A, Jackson T (2016) Understanding households as drivers of carbon emissions. In: Springer ebooks, pp 181–203. https://doi.org/10.1007/978-3-319-20571-7_9
12. Osei MB, Acquaye A, Stamati T (2023) Improving sustainable supply chain performance through organisational culture: a competing values framework approach. J Purchas Supply Manage 29(2):100821. https://doi.org/10.1016/j.pursup.2023.100821
13. Taplin IM (2014) Global commodity chains and fast fashion: how the apparel industry continues to re-invent itself. Compet Change 18(3):246–264. https://doi.org/10.1179/1024529414z.00000000059
14. Wang Y (2011) Consumer behavior characteristics in fast fashion
15. Cachon GP, Swinney R (2011) The value of fast fashion: quick response, enhanced design, and strategic consumer behaviour. Management Science 57(4):778–795. https://doi.org/10.1287/mnsc.1100.1303

16. Engel JF, Blackwell RD, Miniard PW (1995) Consumer behaviour. Dryden Press, United Kingdom
17. Barrera T (2021, November 19) Fashion seasons explained: how it started & how it's *going in 2022 and beyond*. The Tech Fashionista. https://thetechfashionista.com/fashion-seasons-explained/
18. Shen B (2014) Sustainable fashion supply chain: lessons from H&M. Sustainability 6(9):6236–6249. https://doi.org/10.3390/su6096236
19. Bhardwaj V, Fairhurst A (2010) Fast fashion: response to changes in the fashion industry. Int Rev Retail, Distrib Cons Res 20(1):165–173. https://doi.org/10.1080/09593960903498300
20. Kering (n.d.) Kerning.com. Kering's participation to the 2023 ChangeNOW summit [Video]. Kering. https://www.kering.com/en/sustainability/
21. De Klerk A (2019, January 30) Paris is aiming to be the sustainable fashion capital of the world. Harper's BAZAAR. https://www.harpersbazaar.com/uk/fashion/fashion-news/a26082611/paris-sustainable-fashion-captial/
22. AFPRelaxnews & By AFPRelaxnews (2022) Copenhagen, Stockholm, Detroit: three fashion capitals in the making. Forbes India. https://www.forbesindia.com/article/lifes/copenhagen-stockholm-detroit-three-fashion-capitals-in-the-making/79683/1
23. McKinsey & Company (2023, March 8) Generative AI: unlocking the future of fashion. https://www.mckinsey.com/industries/retail/our-insights/generative-ai-unlocking-the-future-of-fashion
24. Vladimirova K, Henninger CE, Armstrong CM, Iran S, Diddi S, Durrani M, Iyer KV, Jestratijevic I, McCormick H, Niinimäki K, Thangavelu P, Sauerwein M, Singh R, Simek P, Wallaschkowski S (2022) Fashion consumption during COVID-19: comparative analysis of changing acquisition practices across nine countries and implications for sustainability. Sciencedirect 5:100056. https://doi.org/10.1016/j.clrc.2022.100056
25. Michel P (2022b, December 31) Less is more: why minimalist fashion is the way of the future. embattled clothing. https://embattledclothing.com/blogs/embattled-academy/less-is-more-why-minimalist-fashion-is-the-way-of-the-future
26. Faux S, Davies P (1988) Wardrobe: develop your style and confidence. Piatkus, United Kingdom
27. Ross E (2021) How to build the best capsule wardrobe? Ask Donna Karan. WWTNT. https://wwtnt.com/how-to-build-the-best-capsule-wardrobe-just-ask-donna-karan/
28. Kim T (2018, April 19) Extreme minimalism—iPhone design philosophy in a word. Medium. https://medium.com/@taehoonkim_22222/extreme-minimalism-iphone-design-philosophy-in-a-word-63c7b29ae497
29. Pangarkar A, Shukla P, Taylor CR (2021) Minimalism in consumption: a typology and brand engagement strategies. J Bus Res 127:167–178. https://doi.org/10.1016/j.jbusres.2021.01.033
30. Kang J, Martinez CMJ, Johnson CL (2021) Minimalism as a sustainable lifestyle: its behavioral representations and contributions to emotional well-being. Sustain Prod Consump 27:802–813. https://doi.org/10.1016/j.spc.2021.02.001
31. Yıldırım S (2023) Greenwashing: a rapid escape from sustainability or a slow transition? LBS J Manage Res 21(1):53–63. https://doi.org/10.1108/lbsjmr-11-2022-0077
32. NielsenIQ (2023, October 4) Consumers care about sustainability—and back it up with their wallets—NIQ. NIQ. https://nielseniq.com/global/en/insights/report/2023/consumers-care-about-sustainability-and-back-it-up-with-their-wallets/
33. Changing Market Foundation (2023, February 16) Frontpage—changing markets. Changing Markets. https://changingmarkets.org/
34. Ranasinghe L, Jayasooriya V (2021) Ecolabelling in textile industry: a review. Resour Environ Sustain 6:100037. https://doi.org/10.1016/j.resenv.2021.100037
35. Allwood JM, Laursen SE, De Rodríguez CM, Bocken NM (2006) Well dressed. The present and future sustainability of clothing and textiles in the United Kingdom, 1

Importance of Market Segmentation and Application of Oekotech

Rena Mehta, Chavi Goyal, and Shalini Gaur

Abstract Internationally eco-textile trade is witnessing a significant surge in demand due to mounting environmental consciousness among consumers. Oekotech textiles, intended with a focus on reducing ecological footprints, are leading the charge in the direction of a sustainable solution for the textile industry. These textiles, made from biological or recycled fibers, minimize the use of detrimental chemicals and promote ethical production practices. The market is divided into various segments, including apparel, home textiles, technical applications, and healthcare, each witnessing a rise in eco-friendly replacements. The future of eco-textile manufacturing looks promising, with a growing number of industries adopting sustainable alternatives. With ongoing innovations and consumer awareness, the industry is poised for continuous growth, offering environmentally conscious choices to consumers across various sectors.

Keywords Ecotech · Sustainable · Oekotech · Manufacturing · Consumer

1 Introduction

An integrated pollution management strategy is necessary to protect our ecosystem from these consequences. Due to unsustainable existing practices, the textile industry has a significant pessimistic influence on the ecosystem. As a result, businesses, environmentalists, and consumers are seeking ways to lessen the textile sector's carbon

R. Mehta (✉)
Department of Fashion Design, Indian Institute of Crafts & Design, Jaipur, India
e-mail: renamehta@iicd.ac.in

C. Goyal
Department of Fashion Technology, NIFT Panchkula, Panchkula, India
e-mail: chavi.goyal@nift.ac.in

S. Gaur
Department of Fashion and Design, Chandigarh University, Mohali, India
e-mail: hod.fashiondesign@cumail.in

Sadhna et al. (eds.), *Climate Action Through Eco-Friendly Textiles*, SDGs and Textiles,
https://doi.org/10.1007/978-981-99-9856-2_9

footprint. As a result, the need for green textiles has grown in recent years. An integrated pollution control strategy is required to protect our environment from these effects. Eco-friendly fabrics can replace conventional products. Recent years have seen a rise in interest in sustainability, particularly in the fashion sector. Brands that have embraced sustainability aren't afraid to boast about it. However, some requirements must be followed for a product, line of apparel, or manufacturing method to be considered truly sustainable [1]. The ecological and social sustainability of a product may be shown by standards and certifications. In today's environmentally conscious world, eco-tech textiles have gained significant attention for their sustainable and innovative qualities. These textiles combine eco-friendly materials, advanced technologies, and functional design to create products that minimize environmental impact while providing superior performance. However, to effectively penetrate the market and cater to the diverse needs of consumers, market segmentation plays a crucial role. By dividing the target market into distinct segments based on various criteria, eco-tech textile manufacturers can tailor their products, marketing strategies, and communication efforts to effectively reach and satisfy specific consumer groups [2].

2 What Are Oekotech and Ecotech?

Oekotech or Eco textiles, also known as sustainable textiles or environmentally friendly textiles, are fabrics and materials that are produced with a focus on reducing their ecological footprint throughout their lifecycle. They aim to minimize the negative environmental impacts associated with traditional textile production, including resource depletion, pollution, and waste generation. Eco textiles encompass various production methods, materials, and certifications that prioritize sustainability, ethical practices, and reduced environmental harm.

The authenticity and credibility of eco textiles are evaluated by certification and standardization. These include GOTS (Global Organic Textile Standard), OEKO-TEX Standard 100, Fair Trade, and Cradle to Cradle certification, etc. These certifications ensure that specific environmental and social criteria are met throughout the textile's lifecycle, from raw material production to manufacturing, and ultimately to the product. Eco textiles encompass a range of sustainable and environmentally friendly fabrics and materials. From organic cotton and hemp to recycled polyester and natural dyes, eco textiles offer alternatives to conventional textiles that prioritize sustainability, reduced environmental impact, and ethical production practices. The adoption of eco textiles contributes to a more sustainable and responsible textile industry [1].

ÖkoTex, also known as OekoTex is a globally recognized certification system for textiles. The ÖkoTex documentation includes diverse classes of products, and individual criteria is applied to each of the selected class. At all dispensation phases, ÖkoTex Standard 100 is a certification that applies to the textile product which incorporates raw materials and finished products. It safeguards the material from any contagious and hazardous elements in it. It is free from substances that are detrimental to human health, such as chemicals, formaldehyde, etc. Step opted by Sustainable Textile Production certification emphasizes justifiable production processes that are sustainable and are executed within the textile industry. It is effective in measuring the parameters in the textile manufacturing process including occupational health and safety, social responsibility, and production of safe products. The certification is completely optional for manufacturers, retailers, and brands. On the other hand, it gives an assurance to the consumer that the products used meet the safety standards and have undergone rigorous testing that complies with the sustainability standards and bears high safety [3].

OekoTex is a quality certification but does not confirm the cent percent sustainability or eco-friendliness of a textile product in all the parameters. Carbon–neutral production, fair trade practices, and recycled materials are some other parameters that contribute to the sustainability of a product. These additional aspects of sustainability can be addressed by other certifications and eco-labels apart from OekoTex. One such initiative is taken ahead by a nonprofit organization known as Better Cotton Initiative (BCI) which promotes the cultivation of sustainable cotton. The NGO works with industry, farmers, stakeholders, and other governmental organizations to attain the goal of the NGO. The NGO provides training to the farmers and educates them about the latest advancements in the field of production in terms of pest control, emphasizing on reduction of wastewater, alternate use of wastewater, and how can one protect the soil. BCI focuses an extensive amount of emphasis on effective irrigation, educating farmers on how to use organic fertilizer and effective irrigation methods. In addition, Farmer training and capacity building provides training courses to help cotton farmers learn new skills and expand their capacities. This includes imparting information and direction on best agricultural practices, environmental stewardship, and workplace health and safety. BCI assists farmers in increasing their output while lowering expenses and having a minimal negative impact on the environment through farmer field schools. The implementation of methods that reduce the damaging effects of cotton cultivation on the environment is encouraged by environmental protection. Among them are initiatives to encourage the preservation of biodiversity, lessen water pollution, and prevent soil erosion [4].

BCI also promotes the preservation of ecosystems impacted by cotton production as well as the protection of valuable habitats. The importance of social and economic development in cotton-producing communities is acknowledged by BCI in its social and economic development criteria. It seeks to enhance the lives of cotton farmers by supporting gender equality, upholding fair labor practices, and encouraging community involvement. BCI aims to create positive social and economic impacts by strengthening farmer organizations and improving access to markets and resources. Collaboration and partnerships foster collaboration with various stakeholders, including cotton farmers, retailers, brands, governments, and NGOs. It fosters partnerships to drive positive change throughout the cotton supply chain. BCI engages with industry players to promote the demand for and uptake of Better Cotton, encouraging market transformation and influencing sustainable cotton sourcing practices. BCI does not have a certification program but operates through a system of mass balance and traceability. Participating farmers and organizations follow the BCI standards and principles, and the better cotton they produce is then mixed with conventional cotton in the supply chain. The World Wide Fund for Nature (WWF) was founded in 1961 and operates in over 100 countries, working toward the conservation of nature and the sustainable use of natural resources. Key aspects included are:

Conservation and Biodiversity: WWF focuses on conserving the world's most ecologically important habitats and species. It works to protect forests, oceans, freshwater ecosystems, and other critical habitats. Conservation WWF promotes management practices of land and water. It works in establishing protected areas and combating wildlife trade.

Change of Climate: WWF is an agency that claims climate change is the most threatening to the planet and its biodiversity. The work of WWF is to restore the wetlands and forests that are considered to be carbon reservoirs on land [5].

Sustainable Development and Environmental Conservation: A balance in environmental conservation is attained by attaining sustainable development. WWF aims at healthy well-being. It works with the local communities and governments to promote sustainable livelihoods along with responsible resource management.

Partnership and Collaboration: WWF leverages collective expertise, resources, and influence to achieve greater conservation impact. WWF's approach to conservation is holistic and multifaceted, combining community-based conservation with education and awareness initiatives to create sustainable solutions that benefit both people and the environment. Through these initiatives, WWF aims to inspire individuals and communities to take action to protect our shared home [6].

The cradle-to-cradle approach to recycling prioritizes the use of safe materials to reduce reliance on nonrenewable resources and minimize the negative impact of waste. Material Reutilization promotes the concept of "Waste Equals Food." It encourages the design of products and systems where all materials used are either biodegradable and returned to the natural environment or can be endlessly recycled into new products without any loss of quality. It encourages the use of clean and renewable energy in all stages of production and manufacturing. Water stewardship emphasizes responsible water management practices, including the efficient use of water resources, the reduction of water pollution, and the protection and restoration of water ecosystems. Social fairness recognizes the importance of social equity and fair treatment of workers throughout the supply chain. It promotes safe working conditions, fair wages, and respect for human rights [7].

Through standardized instruments referred to as the High Index, they assess the sustainability performance of textile firms. bluesign® certifies based on the usage of chemicals, materials, processes, and final textiles are assessed by five sustainability criteria using the bluesign® certification system. Any business, including brands, producers, and chemical suppliers, is referred to be a "bluesign® system partner" if their goods match these requirements. The Organic Content Standard (OCS) is a certification that verifies the percentage of final goods made from organically farmed materials.

3 Market Division for Eco Textile Based on Products

The market division for eco textiles can be categorized based on various factors, including the type of eco textile, end-use applications, and geographical regions. Below are mentioned some common market divisions for eco textiles [8].

3.1 Types of Eco Textiles

Wool and silk have many advantages for consumers in addition to using environmentally friendly production techniques. Wool is a safe material to use for clothes and bedding since it is naturally flame-resistant. Additionally, it is a great insulator, keeping you cool in the summer and warm in the winter. The sumptuous feel and hypoallergenic properties of silk, on the other hand, are well recognized. It is the best material for sleepwear since it is soft on the skin and helps control body temperature. Wool and silk are excellent alternatives for individuals with sensitive skin or allergies since they both have natural antibacterial characteristics that help reduce odors and bacterial growth. Overall, these natural animal fibers offer a sustainable and comfortable option for textiles that prioritize both environmental responsibility and consumer well-being.

3.2 Application of Eco Textiles

Eco textiles are fabrics and materials produced using sustainable and environmentally friendly processes, divided into fashion, home, technical, and nonwoven. Demand for eco textiles is expected to grow as consumers become more aware of the environment [9].

3.3 Geographical Regions

Market divisions are Latin America, the Middle East, Asia Pacific Europe, and North America based on the geographical regions. These divisions help analyze regional trends, consumer preferences, regulatory frameworks, and market dynamics specific to each region.

Others: This segment includes eco textiles used in niche or specialized applications, such as eco-friendly packaging materials, agricultural textiles, and outdoor textiles.

3.4 Global Market Positioning for Oekotech and Eco Textiles

The Global Oekotech Textiles Market in North America, Latin America, Europe, Asia Pacific, and the Middle East and Africa is divided into five regions based on geography. In 2021, the main market for Oekotech textiles will be North America. Major contributors to this amount included the United States, which was followed by Canada, Mexico, and China. This region has shown significant growth over the past few years due to higher sales from new product launches as well as increasing demand arising out of new environmental legislation stipulating that textile products need to be eco-friendly. Latin America showed robust growth rates owing to high economic development status coupled with increased awareness among people about environmental protection along with growing production capacity which are some of the key factors driving this trend. Europe is also a major market for Oekotech textiles.

The region has been the leading supplier of Oekotech products in Europe and accounted for around 27% of total sales volume. The Asia Pacific showed significant growth rates owing to increasing demand arising out of new environmental legislation stipulating that textile products need to be eco-friendly as well as high economic development status coupled with increased awareness among people about environment protection which are some of the key factors driving this trend. China, Australia, India, Japan, and Thailand were some prominent countries generating revenues from the OEKT application in the Asia Pacific. The Middle East and Africa generated $290 million worth of revenue from these applications. Saudi Arabia, Algeria, and South Africa were some major countries contributing to the growth of this region [10].

3.5 Understanding the Market

According to shared traits, tastes, and behaviors, a market can be segmented by breaking it up into more manageable, homogeneous groups. Businesses can adjust their products and marketing strategies to the unique requirements and preferences of various customer groups by segmenting the market to identify these wants and preferences. As a result, firms become more successful and become market leaders in the environmentally friendly sector.

4 Market Segmentation of Eco Textiles

Eco-tech textile businesses can create focused marketing efforts that successfully reach their target market by understanding the various demands and preferences of their clients. For instance, they can develop language that highlights the environmental advantages of their products for consumers who are environmentally sensitive or that highlights the benefits of performance and durability for athletes and outdoor enthusiasts. Market segmentation can also assist businesses in finding untapped client niches or developing market trends that could lead to new chances for growth. Eco-tech textile businesses can establish themselves as pioneers in the sustainable fashion industry by staying on top of trends and adapting to changing consumer demands. In the end, market segmentation is an essential strategy for any company hoping to thrive in the cutthroat business environment of today, and eco-tech textile businesses are no exception. By leveraging this strategy effectively, they can build strong relationships with customers, drive sales growth, and make a positive impact on the environment at the same time. Market segmentation for eco-tech textiles is evident. Based on type, the market is categorized into four categories: woven fabric, Knitted, nonwoven fabric, and others.

Woven Fabric: When it comes to the worldwide market for Ecotech textiles, woven fabric has the biggest market share. The primary reason for this tendency is that end consumers prefer woven materials over nonwoven and knitted fabrics because they provide superior protection. Nonwoven fabric is simply a flat cloth that has been woven by entwining fiber strands, either randomly or in an organized manner. Continuous filaments like viscose rayon, polyester, and nylon are used to make nonwoven textiles. Short fibers with acceptable sewing capabilities but little ability to form high-quality yarns may be used to create economically appealing nonwovens. Knitted fabric is mostly used for applications, clothing, and textiles in the worldwide market for Ecotech textiles. Usually, knitted textiles are used to make sweaters, scarves, gloves, and caps. Given the growing knowledge of health difficulties that may be avoided by wearing wool-enriched clothing, such as the decreased risk of allergies, the market for knitted fabric is predicted to rise over time. Because of expanded application potential, this product area has also seen explosive expansion.

4.1 Factors Affecting the Market Segmentation

Several aspects need to be taken into account to classify the market for eco-tech textiles effectively. Companies can identify particular age groups, genders, income levels, and other significant demographic factors by using demographic segmentation [11]. For instance, younger consumers may be more interested in eco-tech fabrics and often have a higher propensity for sustainable processes. The values, beliefs, interests, and lifestyles of the consumer are the primary subject matter of psychographic segmentation. In this context, people who are concerned about the environment, and their health, or who are interested in technology are possible target markets for eco-tech textiles. Regional preferences and the accessibility of sustainable textile resources are taken into account during geographic segmentation. For example, organizations may concentrate their efforts in regions that place a high priority on environmental protection or in areas where eco-tech textile production may make use of locally obtained resources. The study of consumer purchase trends, brand loyalty, and usage patterns is known as behavioral segmentation. For eco-tech textile businesses, it can be useful to target consumers who already have shown affection for eco-friendly goods or who are early adopters of new technology. Eco-tech textile producers can modify their product lineups and marketing strategies to fit the specific requirements and preferences of each category once the market has been segmented into numerous groups. Businesses may produce sustainable textiles that address environmental concerns while providing efficient and fashionable solutions by developing an awareness of the particular needs of various consumer groups. The use of recycled materials in eco-tech textiles, for instance, might be valued by one market group, while energy efficiency in the manufacturing process might be appreciated by another.

4.2 Benefits of Market Segmentation

In the eco-tech textile sector, market segmentation is the process of breaking a diverse marketplace into easier-to-manage, homogeneous categories based on shared consumer traits, demands, preferences, and behaviors. It enables businesses to recognize specific customer segments with particular demands in terms of pricing, performance, aesthetics, or sustainability. Additionally, it enables businesses to design efficient marketing plans that are specific to each target market segment and customize their goods and services to meet the interests of various consumer categories. A sector interested in eco-tech textiles, for instance, may respond favorably to social media campaigns emphasizing these benefits, whereas a segment interested in luxury may prefer customized advertisements in high-end fashion magazines. Eco-tech textile producers can better target their marketing efforts to maximize their impact and communicate with their target audience by being aware of the particular requirements and preferences of each segment. Manufacturers of eco-tech textiles may additionally use data analytics to monitor consumer behavior and modify their marketing plans as appropriate. In a market that is constantly evolving, this might help them stay one step ahead of the competition and remain relevant. Ultimately, eco-tech textile producers can develop trusting relationships with their clients and encourage steadfast loyalty by adopting a customer-centric marketing strategy. As a result, the fashion industry as a whole may see a rise in sales, better brand awareness, and a more sustainable future.

5 Importance of Oeko Textile and Eco Textile

By allowing customers the option to make more sustainable choices Oekotech Textiles is paving the path for the establishment of a more sustainable and socially responsible textile industry. Consumer desire and awareness: Oekotech textiles meet this need by allowing customers to choose more environmentally friendly clothes and home textiles. This rising demand fosters innovation in eco-friendly textile production and encourages the sector to adopt more sustainable procedures. Demand for sustainable and socially conscious products is increasing as consumers become more conscious of the impact that their purchases have on the environment. Leading this movement, Oekotech Textiles gives customers the chance to select clothes and household textiles that are more environmentally friendly. The industry is moving toward adopting more sustainable practices including using recycled materials and using less water during production as a result of the rising demand for environmentally friendly products. As organizations work to satisfy the demands of customers who care about the environment, it also promotes innovation in the manufacturing of environmentally friendly textiles. Customers can actively contribute to the development of a more just and socially conscious textile sector that places a high value on sustainability and ethical business practices by purchasing Oekotech textiles. Health and wellbeing: Oekotech fabrics can help create a safer and healthier atmosphere at home. They contribute to the creation of a healthier indoor atmosphere, particularly for those with respiratory conditions or chemical sensitivity.

Research, Innovation, and Collaboration: The focus on Oekotech textiles promotes cooperation, research, and innovation within the textile sector. It encourages researchers, designers, and manufacturers to look into innovative, more eco-friendly, efficient, and sustainable technologies, materials, and processes. As a result, the industry is encouraged to continually enhance and move toward higher sustainability [12].

The entire environmental impact can also be influenced by elements including transportation, dyeing procedures, and customer behavior. Eco textiles often have various advantages over regular textiles in terms of the environment. The use of natural, organic, or recycled fibers—which require fewer synthetic chemicals, insecticides, and fertilizers during cultivation—is frequently prioritized in eco textiles, which reduce chemical consumption. As a result, less dangerous compounds are released into the environment, soil and water pollution is reduced, and ecosystems are supported. It leads to the preservation of resources. Eco textiles encourage resource preservation by reducing raw material extraction, energy use, and water use. For instance, growing cotton organically typically uses less water than growing cotton traditionally. Comparably, recycled textiles minimize the demand for virgin resource extraction. Due to the use of recycled materials in processing that reduce waste formation, eco textile also contributes to waste reduction. Eco textiles support waste reduction and a more circular economy by diverting trash from landfills and incineration. Energy-efficient manufacturing techniques are frequently prioritized in eco textiles.

Eco textiles could emphasize water-wise manufacturing techniques and prudent water management. This involves using water-saving methods throughout the dyeing, finishing, and other manufacturing processes. Eco textiles may also help with initiatives to preserve or restore aquatic environments, which would further improve their environmental impact. In comparison to synthetic textiles, some eco textiles, especially those created from natural fibers or recycled materials, offer better biodegradability or recyclability. It suggests that they have a smaller environmental impact at the end of their lifecycle since they either naturally decompose without leaving harmful residues or are recycled into new goods. Low carbon footprints from production and shipping are a common goal of eco textiles. This can be accomplished in several ways, including by using sustainable farming techniques, increasing production efficiency, and procuring supplies locally to cut down on transportation costs.

6 Area and Application of Eco Textiles

The eco-textile market includes some specialties and industries [13]. The following are some critical segments of the eco-textile market:

6.1. **Apparel and Fashion**: One of the market's main sectors for eco textiles is the apparel and fashion industry. Since a growing number of environmentally conscious customers are looking for sustainable solutions, eco-fabrics are frequently employed in the clothing and fashion industries to provide sustainable and ethical garments. These companies can lessen their carbon footprint and generate waste by utilizing green fabrics. For instance, organic cotton is grown without the use of potentially harmful chemicals and pesticides, which is good for the environment and the working conditions of farmers. Less plastic waste ends up in landfills when recycled polyester is made from post-consumer plastic bottles. Compared to traditional cotton, hemp, and linen are much more durable and use less water to grow. Fast-growing bamboo may be harvested sustainably and without endangering the environment. Sustainable fashion manufacturers frequently put ethical labor practices first by guaranteeing fair wages and secure working conditions for their employees, in addition to using eco-friendly textiles. Customers can support ecologically friendly practices while still dressing stylishly by selecting sustainable fashion solutions [14].

Consumers are looking for environmentally appropriate substitutes for traditional textiles for their homes, such as organic or recycled materials, sustainable manufacturing techniques, and certifications like GOTS. Eco textiles are becoming more and more well-liked in the home textiles and furnishings business as consumers grow more conscious of how their decisions affect the environment. The products, which range from eco-friendly bedding to upholstery fabrics and curtains, have several advantages that go beyond just being green. They are free of hazardous chemicals, made from organic or recycled materials, and produced using environmentally friendly production techniques. This essentially means that they are beneficial for both the environment and our health. Making eco-friendly decisions for our home textiles and furnishings is now simpler than ever thanks to the variety of options available, from bathrobes to rugs and carpets. So, while shopping for a cozy new blanket or a chic new rug, think about selecting an eco-textile alternative that will not only look wonderful in your house but also contribute to protecting the environment for future generations.

Technical textiles are used in industrial and technical applications, including eco textiles, which offer durability, strength, and functional properties while minimizing environmental impact.

Nonwoven fabrics are frequently utilized in a variety of products, including disposable hygiene items, medical textiles, geotextiles, and more. In this industry, eco textiles composed of recycled materials or biodegradable fibers have gained popularity.

Eco fabrics are becoming more and more appreciated in the sports and outdoor gear industry, providing athletes and outdoor enthusiasts usefulness, durability, and benefits for the environment. Compared to conventional materials, they also provide better performance and durability. Eco textiles are growing in popularity across industries, which is not surprising given the increase in conscientious consumerism. We may anticipate a change toward an eco-friendlier future as more businesses adopt sustainable practices [4].

The use of eco-fabrics in fashion and apparel has grown in popularity as designers and brands incorporate sustainable materials into the designs they create. Cotton, hemp, bamboo, and Tencel are just a few of the organic or recycled materials used to make these fabrics. They offer a more sustainable alternative to conventional fabrics, which are usually created with risky chemicals and pollute the environment. The textile industry's usage of water and carbon emissions may be reduced with the use of eco textiles. Furthermore, sustainable textile production may strengthen local economies by supporting local farmers and craftspeople. It is projected that as people grow more aware of how their purchases affect the environment, demand for eco-friendly products will rise encouraging the use of eco-friendly textiles in various industries [15].

As consumers become more aware of how they impact the environment and the demand for sustainable products rises, the worldwide eco-textile industry is anticipated to continue expanding, presenting a business opportunity for textile producers and retailers.

6.2. **Medical and Healthcare**: Sustainable medical textiles, surgical gowns, masks, wound dressings, and bedding for hospitals and healthcare facilities are made using eco textiles in the healthcare and medical sectors. To ensure patient comfort and minimal environmental impact, these materials are made to be secure, hypoallergenic, and green.

Ecotech is in high demand, and ecological factors such as growing awareness of climate change and the need for resource conservation affect how customers behave. For the market to expand and endure, a reliable supply network for these commodities is essential. The eco-tech market may be impacted by legal concerns such as trademarks and intellectual property rights. To protect their ideas from infringement, businesses that invest in the development of novel manufacturing techniques or sustainable technologies need legal protection. Manufacturers of Oekotech textiles are required to adhere to statutory requirements as well as industry standards for product safety, labor rights, and sustainability. Noncompliance may harm one's reputation and legal repercussions. Oekotech textile producers may better comprehend the external environment, spot opportunities, reduce risks, and match their plans with market expectations by taking these PESTEL aspects into account the examination of market competition dynamics [8].

Due to investment in R&D, eco-friendly production techniques, and supply chains for sustainable materials, the Oekotech textiles market may be significantly challenged by new entrants. Businesses can reduce the risks associated with relying on a single source of sustainable materials by working to diversify their supply base. By implementing these actions, businesses may contribute to the development of a more sustainable supply chain overall and guarantee that they have access to the sustainable products they require. Although there may be replacement sustainable materials or methods, their overall threat is still minimal.

7 Conclusion

By minimizing waste and decreasing chemical use, Ecotech materials are essential for promoting an eco-friendly atmosphere. It helps in adhering to moral standards that would satisfy the rising demand for environmentally friendly textiles. They provide buyers with the opportunity to choose ecologically friendly options while supporting another ethical and conscientious textile sector. Adopting certified Oekotech textiles can help create a future that is more sustainable and environmentally friendly. One of the factors boosting this market's demand over the past few years has been the rise in customer desire for safety and hygiene. As a result, top manufacturers like Nike Inc., Adidas AG, Under Armour Inc., etc., have introduced their products in this niche. Businesses may successfully explain the value proposition of their sustainable textile goods and obtain a competitive advantage in the market by finding and focusing on niche consumer segments. Market segmentation also enables businesses to properly manage resources, improve customer happiness, and profit from new trends. Ecotech textile companies must establish a market segmentation strategy to succeed in this competitive industry as demand for sustainable textiles increases. Oekotech textile producers face obstacles in the market, including competitiveness, access to sustainable supplies, and customer expectations. To succeed in this changing market, companies will need to continuously innovate, differentiate themselves, and manage their supply chains efficiently.

References

1. Adeel S, Kamal S, Ahmad T, Bibi I, Rehman S, Kamal A, Saleem A (2020) Biotechnology: an ecofriendly tool of nature for textile industries. In: Advances in functional finishing of textiles, pp 85–114
2. Aishwariya (2022) Inclination towards OekoTextiles. Chief Editor Dr. R. Jaya Kumar, pp 155
3. Bhamra T (2007) Building ecodesign throughout the supply chain: a new imperative for the textile and clothing industry. Woodhead Publishing, pp 41–49
4. Karthik T, Gopalakrishnan D (2014) Environmental analysis of textile value chain: an overview. Roadmap to sustainable textiles and clothing: environmental and social aspects of textiles and clothing supply chain, pp 153–188
5. Blackburn R (ed) (2009) Sustainable textiles: lifecycle and environmental impact. Elsevier
6. Lara L, Cabral I, Cunha J (2022) Ecological approaches to textile dyeing: a review. Sustainability 14(14):8353
7. Kumari P, Singh SS, Rose NM (2013) Eco textiles: for sustainable development
8. Suparna MG, Rinsey Antony VA (2016) Eco-friendly textiles. Int J Sci Technol Manage 5(11):67–73
9. Nemcic I (2017) Importance of Oeko tex standard for textile industry. Doctoral dissertation, University of Zagreb. Faculty of Textile Technology
10. Rungruangkitkrai N, Mongkholrattanasit R (2012) Ecofriendly of textiles dyeing and printing with natural dyes. In: RMUTP international conference: textiles and fashion. 3(4):1–17
11. Sewekow U (1996) How to meet the requirements for Eco textiles. Textile Chem Colorist 28(1)
12. Patnaik A (2019) Technical textiles and recent developments. Fiber to smart textiles, pp 315–330
13. Miraftab M, Horrocks AR (eds) (2007) Ecotextiles: the way forward for sustainable development in textiles. Elsevier.
14. Gardetti MA, Torres AL (eds) (2017) Sustainability in fashion and textiles: values, design, production and consumption. Routledge
15. Maity S, Singha K, Pandit P (2023) Introduction to functional and technical textiles. In: Functional and technical textiles. Woodhead Publishing, pp 1–30

Implementation and Environment Protection Through ISWM (Integrated Solid Waste Management) in Textiles

J. Jaisri and S. Balaji

Abstract In the ever-evolving realm of textile production, the need for sustainable practices and environmental consciousness has become paramount. This abstract delves into the exploration of Integrated Solid Waste Management (ISWM) within the textile industry, aiming not just for theoretical understanding but a practical roadmap for implementation and robust environmental protection. The chapter unfolds against the backdrop of a textile industry facing growing challenges in waste management. By navigating the landscape of ISWM, it seeks to unravel innovative strategies and holistic approaches for handling solid waste generated in textile processes. The emphasis is not only on mitigating the environmental impact but also on the effective implementation of these strategies within the industry's operational framework. Key components of the abstract include an analysis of the current state of solid waste generation in textile manufacturing, the identification of sustainable waste management practices, and a comprehensive exploration of integrated approaches that synergize environmental protection with operational efficiency. As the textile industry stands at the crossroads of innovation and environmental responsibility, this chapter serves as a guide for stakeholders, researchers, and practitioners alike. It envisions a future where the threads of sustainable textile production are intricately woven into the fabric of environmental stewardship.

Keywords Textile · Sustainable · Waste management

J. Jaisri (✉)
Department of Costume Design and Fashion, Nehru Arts and Science College, Coimbatore 641 105, India
e-mail: nascjjaisricdf@nehrucolleges.com

S. Balaji
School of Commerce, Nehru Arts and Science College, Coimbatore 641 105, India

Sadhna et al. (eds.), *Climate Action Through Eco-Friendly Textiles*, SDGs and Textiles,
https://doi.org/10.1007/978-981-99-9856-2_10

1 Introduction

One of the most significant and ancient industries in India is the garment business. It has a long history that dates back to ancient times, and it continues to play an important part in our nation's industrial and economic landscape. The textile business includes textile design, development, production, and distribution in addition to textiles and fabrics [1].

Before the modern era, individuals made clothing and textiles themselves for their personal use. On some occasions, they were simply resold. The invention of the flying shuttle in 1733, the spinning jenny in 1764, and the power loom in 1784 marked the beginning of the clothing industry. Mass production of materials and apparel began after that. The improvements to James Watt's steam engine in 1775, Elias Howe's sewing machine in 1846, and Eli Whitney's cotton gin in 1792 all contributed to the textile industry's rapid expansion.

A major employer in India is the clothing industry, employing a vast number of people in both urban and rural regions. It provides a living for weavers, artisans, agriculturalists, and factory employees. Along with other significant industries, our economy is heavily reliant on textile production and commerce. Textile and apparel exports alone account for around 27% of foreign exchange earnings [2]. The textile and apparel industry makes up 3% of the nation's GDP and about 14% of industrial production. The garment sector provides about 8% of the total excise tax collection. In fact, to the point, up to 21% of the employment opportunities in the economic system are in the fashion industry. Roughly thirty-five million individuals work directly in the industry of textiles. More than sixty million individuals may have jobs that are indirect due to the manufacturing of agriculturally related raw materials like cotton along with trade and manufacturing.

Textiles stand as India's largest standalone industry, contributing to over 20% of the total manufacturing output. It directly employs nearly 20 million workers. The exports of textiles and clothing constitute thirty percent of the country's total export value [3]. With 1227 textile mills, India has a total spinning capacity of about 29 million spindles. Although mills produce most of the yarn, the power loom and handloom industries also produce textiles [4]. Cotton remains the dominant raw material in the Indian textile industry, accounting for around 65% of its usage. Cotton cloth production amounts to about 12.8 billion meters (42 billion feet) annually. Jute products contribute the second-highest in global production at 1.1 million metric tonnes. One of the oldest industries in India, the textile sector employs millions of people in the organized, decentralized, and household sectors as well as contributes over 14% to manufacturing value-addition and nearly one-third of gross export earnings. It also includes cotton and jute growers, craftspeople, and weavers.

1.1 The Formation and Expansion of the Indian Textile Industry

The industry related to textiles in India is like a vibrant tapestry, woven with threads of tradition and innovation. Structurally, it's quite diverse, encompassing various segments like cotton, silk, wool, and synthetic fibers. Let's unfold this fabric a bit [3].

1. **Cotton Dominance**: Cotton holds a dominant position, forming a substantial part of the industry, and India stands as one of the leading global producers of cotton.
2. **Spinning Mills**: At the heart of the textile web are spinning mills. They convert raw cotton fibers into yarn, which is the foundation for further processing.
3. **Weaving and Knitting**: Weaving and knitting units take the yarn and create the actual fabric. India has a rich heritage of handloom and power loom weaving, contributing to the diversity of textiles.
4. **Dyeing and Printing**: Once the fabric is ready, it goes through dyeing and printing processes to add color and patterns. Traditional block printing and tie-dye methods still coexist with modern printing techniques.
5. **Garment Manufacturing**: Garment manufacturing is a key sector, transforming fabrics into finished products. From traditional attire to contemporary fashion, this segment caters to diverse tastes.
6. **Synthetic Fibers**: Alongside natural fibers, there's a growing presence of synthetic fibers like polyester and nylon. These materials offer durability and versatility [3].
7. **Silk and Wool**: The industry embraces the luxurious silk and warm wool sectors. India is renowned for its silk sarees, and wool finds its place in winter wear and carpets.
8. **Handicrafts and Handlooms**: Handicrafts and handloom industries add a unique touch. They showcase the skill and artistry of Indian craftsmen, connecting the industry with cultural roots [5].

Growth Drivers

- **Exports**: The sector maintains a notable presence in the global market, exporting Indian textiles to diverse nations.
- **Technology Adoption**: Integration of technology enhances efficiency and quality in production processes.
- **Government Initiatives**: Policies and schemes supporting the textile sector contribute to its growth.
- **Innovation**: The adoption of sustainable practices and innovative designs keeps the industry competitive.

Challenges

- **Global Competition**: The industry faces stiff competition globally.
- **Supply Chain Issues**: Disruptions in the supply chain can impact production.

- **Environmental Concerns**: Balancing growth with sustainability is a challenge.

Future Outlook

- The future looks promising with a focus on sustainable practices, technological advancements, and market diversification.
- In essence, the Indian textile industry is a kaleidoscope of tradition, innovation, and economic significance, weaving together the past, present, and future [5].

1.1.1 Growth of Textile Industry

The growth of the textile industry is a fascinating journey marked by evolution, challenges, and innovation. Over the years, it has woven its narrative into the economic and cultural fabric of nations. Let's unravel the threads of this growth [6].

1. **Historical Tapestry**: The textile industry's growth is deeply rooted in history, tracing back to the earliest forms of textile production. From handloom weaving to the advent of industrialization, the industry has continually transformed.
2. **Industrial Revolution Impact**: The Industrial Revolution acted as a catalyst, propelling the textile sector into mechanized production. Factories emerged and spinning and weaving processes underwent significant advancements.
3. **Global Expansion**: With globalization, the industry's footprint expanded globally. Different regions became specialized in specific types of textiles, fostering international trade and collaboration.
4. **Technological Innovations**: The growth narrative is interwoven with technological innovations. Automation, computer-aided design, and advanced machinery have revolutionized production processes, enhancing efficiency and quality [5].
5. **Market Diversity**: The industry's growth is mirrored in the diversity of its markets. From traditional textiles deeply rooted in cultural heritage to high-tech fabrics meeting contemporary demands, the market has evolved to cater to a broad spectrum of consumer needs.
6. **Sustainable Practices**: In the recent past, there has been an increasing trend of emphasis on sustainable practices. The industry is increasingly adopting eco-friendly materials, ethical manufacturing processes, and circular economy principles, aligning growth with environmental responsibility.
7. **Challenges and Resilience**: Challenges such as global competition, supply chain disruptions, and changing consumer preferences have tested the industry's resilience. Yet, it continues to adapt, demonstrating its ability to navigate uncertainties [3].
8. **Economic Contribution**: The textile industry stands as a significant contributor to the economic tapestry of nations. It provides employment opportunities, contributes to GDP, and plays a crucial role in trade balances [6].
9. **Future Threads**: Looking ahead, the industry's growth is anticipated to be shaped by factors such as digitalization, artificial intelligence, and the pursuit of sustainable and ethical practices. The future promises a dynamic interweaving of tradition and innovation.

In essence, the growth of the textile industry is a story woven with threads of progress, adaptation, and a continuous quest for excellence in craftsmanship and technology [5].

2 The Scale of the Textile Industry in India

The magnitude of India's textile industry is vast and encompasses a broad spectrum of activities, reflecting its significance in the nation's economic landscape. Let's explore the dimensions that define the size of this dynamic industry:

1. **Diverse Activities**: The mining and processing of basic components like cotton, jute, silk, and wool as well as the creation of high-value goods like fabrics and clothing constitute every part of India's textile sector.
2. **Fiber Variety**: The textile industry in India uses a broad range of fibers, including synthetic fibers like polyester, viscose, and acrylic, and blends of these fibers as well as filament yarn and natural fibers like cotton, jute, silk, and wool.
3. **Economic Contribution**: An important part of the Indian economy is the textile sector. It contributes significantly to industrial production, accounting for 20 percent of it. Additionally, it plays a crucial role in excise collections, employment, export earnings, and the overall Gross Domestic Product (GDP).
4. **Employment Generation**: An estimated 35 million people are employed directly by the industry, making it an important source of employment. Given that it is the nation's second-largest employer, highlighting its socio-economic impact [6].
5. **Export Dynamics**: India exports a significant number of textiles, making up close to 20% of all export revenue. One important measure of the sector's size and impact is how well it performs internationally.
6. **Growth Trends**: Over the years, the industry has showcased growth trends, with periods of acceleration, particularly influenced by policy interventions. With an emphasis on economic liberalisation, the Economic Policy of 1991 and the Textile Policy of 1985, played a significant role in accelerating growth during the 1990s.
7. **Global Standing**: India holds a significant position globally in the textile trade. Its textiles are sought after in international markets, further amplifying the size and impact of the industry.

In summary, the textile industry in India is characterized by its sheer scale, encompassing a multitude of activities, contributing substantially to the economy, and serving as a powerhouse of employment. Its size is not only measured in economic terms but also in its intricate interweaving with the social and cultural fabric of the nation.

3 The Economic Importance of the Indian Textile Industry

India's textile industry plays a vital role in the nation's economy, making numerous, large-scale contributions [6]. Let's unravel the intricate threads that define its role:

1. **Economic Contribution**: The textile industry is a formidable contributor to India's economic tapestry. It adds significant value to industrial production, accounting for 20 percent of the total. This contribution underscores its central role in shaping the economic landscape [7].
2. **Excise Collections**: Beyond production, the industry is a key player in government revenues, constituting 9 percent of excise collections. Its economic significance is mirrored in the financial contributions it makes to the national exchequer.
3. **Employment Generation**: With a workforce of approximately 35 million, the textile sector emerges as a major employer in the country. Its role extends beyond economic parameters, impacting the lives and livelihoods of a substantial segment of the population.
4. **Export Powerhouse**: The industry holds the reins of India's global trade prowess, contributing approximately 20% of the nation's overall export revenue. Its products are sought after in international markets, showcasing its global standing [6].
5. **GDP Booster**: The textile sector's impact resonates in the Gross Domestic Product (GDP), contributing a significant 4 percent to the nation's economic output. This places it as a critical driver of economic growth and stability.
6. **Rural and Semi-Urban Linkages**: The textile sector has direct ties to the rural economy in addition to metropolitan settings. Its productivity is related to important fiber-producing crops and industries like cotton, wool, silk, and handlooms. In regions that are semi-urban or rural, this connectivity supports thousands of producers and craftspeople through a cascade effect.
7. **Social Significance**: The industry's role extends beyond economic metrics; it carries cultural and social weight. Traditionally significant crafts like handlooms and handicrafts find expression within the textile sector, contributing to the preservation of cultural heritage [7].
8. **Growth Catalyst**: Policy measures frequently have an impact on the Indian garment sector's economic trajectory. Proposals like the Economic Policy of 1991 and the Textiles Policy of 1985, which prioritized liberalization, served as impetuses that propelled the sector into faster growth stages.

In essence, the Indian textile industry isn't merely a sector; it's a dynamic force shaping the economic, social, and cultural fabric of the nation. Its role spans from the intricacies of craftsmanship to the macroeconomic dynamics of global trade, weaving together a narrative of resilience, growth, and multifaceted significance.

4 Scenario of Textile Industry

The current scenario of the textile industry is a dynamic tapestry shaped by a combination of challenges, innovations, and global trends. Let's unfold the current threads of this landscape [3].

1. **Global Supply Chain Dynamics**: The industry operates in a world where supply chains are interconnected and influenced by global events. The risk of these has been brought to light by recent disruptions like COVID-19 which affects distribution and manufacturing.
2. **Sustainable Practices**: Sustainability is becoming more and more significant in the textile industry. Consumer awareness of eco-friendly products is driving the industry to implement sustainable practices, such as the use of organic materials, moral production methods, and waste minimization programs [7].
3. **Technology Integration**: An important factor in the industry's transformation is technology. Technology is increasing productivity, cutting costs, and fostering innovation in a variety of fields, from automation in manufacturing processes to the application of artificial intelligence in design and production.
4. **E-Commerce Boom**: Consumer access to textiles has changed as a result of the growth of e-commerce. Online marketplaces have grown to be important channels for the sale of clothing and textiles, changing consumer behavior and presenting the sector with both new opportunities and challenges [3].
5. **Changing Consumer Preferences**: Consumer tastes are changing, with an emphasis on ethical and conscientious purchasing. The increasing need for customized, one-of-a-kind, and environmentally friendly textiles is impacting industry participants' tactics.
6. **Policy and Trade Dynamics**: Trade agreements and policies have an impact on the sector. The competitive landscape can be shaped by shifts in trade dynamics, tariffs, and international relations, which can affect the flow of finished goods and raw materials [6].
7. **Innovation in Materials**: The industry is seeing a revolution in materials as sophisticated textiles with novel functions are developed. For instance, smart textiles incorporate technology to offer functions like temperature control and monitoring.
8. **Circular Economy Initiatives**: There's a growing emphasis on circular economy principles within the industry. Efforts to reduce waste, recycle materials, and adopt closed-loop systems are gaining momentum as part of a broader commitment to environmental stewardship.
9. **Resilience Amid Challenges**: Challenges such as fluctuating raw material prices, global economic uncertainties, and competition from low-cost manufacturing centers pose ongoing challenges. The industry's resilience is tested as it adapts to navigate these uncertainties.

In essence, the current scenario of the textile industry is characterized by a delicate interplay of global and local factors, technological advancements, and a shifting consumer landscape. Navigating this landscape requires a strategic blend of innovation, sustainability, and adaptability.

5 Textile Wastage

Materials that are still available after a process or product has served its original purpose are referred to as waste. It is a particular kind of useless, flawed, or ineffectual substance. Everything that is left behind after a particular fabric product has been made but has stopped being valued or usable is called textile waste. During the whole textile production process—spinning, weaving, knitting, dying, finishing, and clothing—waste is produced [8].

5.1 Types of Textile Waste

Spinning, weaving, dying, finishing, the production of clothing, or the customer end of the supply chain are all possible sources of textile waste. In the parts that follow, we'll discover more about the specifics.

1. **Spinning Waste**: Cotton fiber bales contain a variety of wastes, such as foreign objects, dust, seeds, short fibers, and others. As these bales move through various stages of a spinning mill, different types of waste are produced in each stage, such as blowroom waste, carding section waste, roving waste, and so on.
2. **Weaving Waste**: Weaving Residue: Similar to spinning mills, weaving machines generate diverse waste. Wastages refer to the leftover strands from the warping process that remain on the cones. It is inevitable to have some yarn left on the cones as it is impractical to empty the warping creel area of all the cones. Sizing waste is an additional sort of waste in a weaving factory. To make sure that the right-sized yarns are put on the weaver's beam when a new batch of warping yarn starts spinning in the weaver's beam section, some skeins must be removed [7].
3. **The sizing waste** is followed by the knotting waste. Two beams of warp ends must be presented for connecting before knotting. Beam leftover material is another sort of weaving trash [3]. Warp yarn of a small amount is left unfinished on the weaver's beam when a weaver beam is finished and cannot be completed. Another sort of weaving waste is supplementary selvedge waste. A fictional selvedge called an auxiliary selvedge.
4. **Knitting Waste**: The history of knitting is rich and lengthy. Knitting can be done on a machine or by hand. Knitting methods and techniques differ greatly. If there is a problem with the raw ingredients or the knitting process, knitting waste will be produced. We now know about the various types of knitting trash. The merchandiser starts by creating a sample whenever a fresh order is created. To produce a sample, the knitting machine is put through trials. Testing results

in the production of knitting waste. When crocheting, yarn might lead to waste on the floor. If there are flaws in both the yarn and the cone, waste may result [3]. Another reason for knitting waste is fly creation from different yarn guides. Knitted fabric defects include thinning, thickening, and spiraling.

5. **Dyeing Waste**: The most frequent sources of wastewater are the textile and dyeing industries, which seriously jeopardize our ecosystem [9]. Numerous machine makers are making an effort to offer innovative wastewater-reduction solutions. Some people are trying to develop coloring techniques without water. In addition, there are numerous kinds of coloring defects. Many different types of dyeing mistakes result in waste. The most frequent dyeing errors include metamerism, crease lines, batch-to-batch shade variation, uneven coloring, and selvedge to selvedge shade variance for denim. Due to these flaws, waste is generated on the dyeing floor [10].
6. **Consumer Waste**: The apparel industry is divided into several segments, including trimming, sewing, the printing process, work of embroidery, and finishing, as well as packaging and shorting. Produce waste in each component. The cutting area is the main source of waste in the clothing manufacturing industry. Because of the several uses and functions of markers, a considerable amount of trash is generated in the cutting zone. All body parts are inspected after cutting and then shorted and wrapped. Consequently, certain defective parts can wind up as waste in this sector. In the sewing area, the conveyors then disperse these wrapped parts [1].

When an employee in the sewing department finds a defective item, they tend to ignore it. Consequently, waste is produced in the stitching department. The garment piece will be thrown away if some of the patterns in the processing section don't match the requirements. The clothing item will be deemed a waste if the embroidered part is not correctly embroidered. There will be a waste if there are any measuring, trimming, or pressing errors in the finishing phase. Over the previous decade, worldwide garment production has more than doubled. The typical life span of a garment product is three years. The average consumer buys 50% more clothing annually and wears it for about half as long as they did fifteen years ago, which leads to enormous amounts of textile waste [10].

6 SWOT ANALYSIS of the Indian Textile Industry

Strengths

- The Indian textile industry benefits from abundant raw materials. India is a leading global producer of cotton and has substantial resources of fibers like polyester, silk, viscose, and others.

- It possesses a significant advantage in terms of a well-trained workforce. The country's lower wage rates provide a substantial competitive edge, naturally reducing manufacturing costs in the textile industry to highly reasonable levels.
- The spinning sector in India is notably competitive and has established a presence across nearly every link in the worth of cycle. The Indian apparel sector displays remarkable diversity in terms of its scale, manufacturing capabilities, types of apparel produced, production volume and quality, costs, and fabric requirements. It includes manufacturers of ready-to-wear items for the domestic and international markets.

Weakness

- Manufacturing garments of knitting remains highly fragmented, posing challenges for global players who prefer consolidating their sourcing with just a couple of suppliers, a capacity requirement that Indian garment units struggle to fulfill.
- The industry continues to grapple with historical regulations, including the persistence of knitted garments within the Small Scale Industry (SSI) domain.
- The labor force in India exhibits lower productivity compared to competing nations.
- Despite initiatives like TUFS, technological obsolescence remains a concern.
- The industry faces limited bargaining power in a market dominated by customers.
- India's lack of membership in crucial trade pacts restricts its access to main marketplaces.
- The fact that Indian labor rules are comparatively unfriendly to trades highlights the pressing need for labor reforms in the nation.

Opportunity

- There appears to be significant space for expansion based on the low per-capita domestic textile consumption.
- The domestic market is highly attuned to fashion trends, driving the evolution of a nimble garment industry [11].
- India holds a mere 3% of the global market share, whereas China commands around 15%. In the years following 2005, it's anticipated that China will seize 43% of the global textile trade.
- The development of novel goods ought to be the main priority.
- There's a growing reliance on CAD for enhancing design capabilities and fostering greater creativity.

Threats

- Post-2005, competition is expected to extend beyond exports and may also emerge within the country due to the arrival of cheaper, higher-quality imports.
- Standards like SA-8000 or WARP have heightened the expectations for companies to enhance their workplace practices. Other competitive advantages are likely to remain as barriers [12].

7 Integrated Solid Waste Management in Textile

The current scenario of the textile industry is a dynamic tapestry shaped by a combination of challenges, innovations, and global trends [1]. Let's unfold the current threads of this landscape:

1. **Global Supply Chain Dynamics**: The industry operates in a world where supply chains are interconnected and influenced by global events. The vulnerability of these supply chains has been brought to light by recent disruptions like the COVID-19 epidemic, which has an effect on distribution and manufacturing [5].
2. **Sustainable Practices**: In the textile sector, sustainability is becoming more and more important. Consumers are increasingly conscious of eco-friendly products, pushing the industry toward adopting sustainable practices, including the use of organic materials, ethical manufacturing processes, and waste reduction initiatives.
3. **Technology Integration**: An important factor in the industry's transformation is technology. Technology is increasing productivity, cutting costs, and fostering innovation in a variety of fields, from automation in manufacturing processes to the application of artificial intelligence in design and production.
4. **E-Commerce Boom**: Consumer access to textiles has changed as a result of the growth of e-commerce. Online marketplaces have grown to be important channels for the sale of clothing and textiles, changing consumer behavior and presenting the sector with both new opportunities and challenges.
5. **Changing Consumer Preferences**: Consumer tastes are changing, with an emphasis on ethical and conscientious purchasing. The increasing need for customized, one-of-a-kind, and environmentally friendly textiles is impacting industry participants' tactics.
6. **Policy and Trade Dynamics**: Trade agreements and policies impact the industry. The flow of raw materials and completed goods can be impacted by changes in trade dynamics, tariffs, and international relations, which can change the competitive environment.
7. **Innovation in Materials**: The industry is seeing a revolution in materials as sophisticated textiles with novel functions are developed. For instance, smart textiles incorporate technology to offer functions like temperature control and monitoring [11].
8. **Circular Economy Initiatives**: There's a growing emphasis on circular economy principles within the industry. Efforts to reduce waste, recycle materials, and adopt closed-loop systems are gaining momentum as part of a broader commitment to environmental stewardship.
9. **Resilience Amid Challenges**: Challenges such as fluctuating raw material prices, global economic uncertainties, and competition from low-cost manufacturing centers pose ongoing challenges. The industry's resilience is tested as it adapts to navigate these uncertainties.

In essence, the current scenario of the textile industry is characterized by a delicate interplay of global and local factors, technological advancements, and a shifting consumer landscape. Navigating this landscape requires a strategic blend of innovation, sustainability, and adaptability.

8 Textile Waste Management

When discussing how to be more environmentally friendly, greener, and ecologically sound, and how to live sustainably, the phrase Reduce, Reuse, Recycle is frequently used.

What does it mean, though?

When it comes to the textile business, which is well recognized for its resource- and pollution-intensive processes, reducing, reusing, and recycling are crucial actions to take.

Here are strategies and practices to promote sustainable development across the apparel sector:

8.1 Reduce

(a) **Environmentally Friendly Materials**: Encourage the use of environmentally friendly materials like recycled fibers, hemp, and organic cotton. These parts use less energy and leave less of an environmental impact.
(b) **Design for Durability**: Create products that are durable and long-lasting. High-quality textiles can reduce the frequency of replacements, which in turn reduces waste.
(c) **Minimalist Fashion**: Promote minimalist and timeless fashion designs to reduce the demand for fast fashion and reduce the overall production of textiles.
(d) **Efficient Manufacturing**: Adopt lean manufacturing practices to reduce waste during production, such as optimizing cutting patterns to minimize fabric waste.

8.2 Reuse

(a) **Secondhand Clothing**: Encourage the purchase and sale of secondhand clothing through thrift stores, online marketplaces, and clothing swaps.
(b) **Upcycling**: Promote creative ways to upcycle old textiles into new products. For example, turning old jeans into tote bags or repurposing t-shirts into quilts.
(c) **Rental and Sharing Platforms**: Support clothing rental and sharing platforms where people can rent garments for special occasions instead of buying new ones.

(d) **Repair and Alteration**: Encourage the use of repair and modifying solutions to increase the longevity of clothes rather than discarding them [3].

8.3 *Recycle*

(a) **Textile Recycling Facilities**: Invest in textile recycling facilities that can break down old textiles into raw materials for new products.
(b) **Collection Programs**: Establish collection programs for old clothing and textiles, making it easy for consumers to dispose of them responsibly.
(c) **Recycled Fiber Production**: Use recycled fibers in the production of new textiles. This reduces the need for virgin materials.
(d) **Closed-Loop Systems**: Implement closed-loop systems where textiles are collected, recycled, and turned into new products within the same supply chain.

Consumer Education

(a) **Raise Awareness**: Inform customers about the advantages of reducing, reusing, and recycling textiles as well as the effects that the textile industry has on the environment [4].
(b) **Care Instructions**: Provide care instructions to help consumers maintain and extend the life of their clothing.

Regulation and Certification

(a) **Support Sustainable Certifications**: To guarantee ethical and ecological practices in the textile sector, and promote the adoption of certifications.
(b) **Regulatory Policies**: Advocate for government policies that promote sustainable practices, such as taxes on non-recyclable textiles or incentives for recycling.

Collaboration

(a) **Industry Collaboration**: Collaborate with other stakeholders, including brands, manufacturers, NGOs, and government agencies, to develop and implement sustainable practices.
(b) **Research and Innovation**: To create new sustainable materials and technologies for the textile industry, allocate resources to research and innovation.

Reducing, reusing, and recycling in the textile industry requires a concerted effort from all stakeholders, from manufacturers and retailers to consumers and policymakers. By adopting these strategies, the textile sector has the potential to greatly lessen its influence on the environment and help create a healthier future [3].

9 Take Action Against Textile Waste

Minimizing the clothing sector's negative environmental impact requires action against textile waste [7]. Here are various methods to address textile waste at different stages of the textile lifecycle:

9.1 Consumer Awareness and Education

- **Raise Awareness**: Educate consumers about the ecological and community ramifications of rapid fashion and excessive textile use.
- **Inform About Sustainable Choices**: Provide information on sustainable fashion choices, such as buying durable, timeless clothing, supporting eco-friendly brands, and adopting a minimalist wardrobe.
- **Teach Repair Skills**: Encourage consumers to learn basic sewing and mending skills to extend the life of their clothing.

9.2 Circular Fashion Initiatives

- **Clothing Rental**: Promote and support clothing rental services where consumers can borrow clothing for a period instead of buying.
- **Swap and Thrift**: Organize clothing swap events and promote thrift store shopping as an alternative to buying new clothes.
- **Take-back programs**: Encourage brands to establish take-back programs, allowing customers to return old clothing for recycling or upcycling.

9.3 Retail and Brand Responsibility

- **Sustainable Manufacturing**: Brands should prioritize sustainable and ethical manufacturing practices, including using eco-friendly materials and reducing waste in the manufacturing process.
- **Objectivity**: Encourage supply chain transparency so that consumers may make educated purchasing decisions.
- **Extended Producer Responsibility (EPR)**: Advocate for EPR programs where brands take responsibility for the disposal and recycling of their products.

9.4 *Support Textile Recycling Facilities*

- Invest in or support the development of textile recycling facilities that can convert old clothing into new materials.
- **Upcycling Initiatives**: Encourage designers and entrepreneurs to create new products from old textiles through upcycling.

9.5 *Government Regulation and Policies*

- **Textile Waste Regulations**: Advocate for policies that restrict the disposal of textiles in landfills and promote recycling and reuse.
- **Tax Incentives**: Lobby for tax incentives for businesses that adopt sustainable practices and use recycled materials.
- **Labeling Requirements**: Push for labeling requirements that disclose information about the environmental impact of clothing items.

9.6 *Support Sustainable Fashion Brands*

- **Choose Sustainable Brands**: Support brands that prioritize sustainability and ethical practices in their supply chain.
- **Local and Artisanal Products**: Explore locally made or artisanal clothing options, which often have a smaller carbon footprint.

9.7 *Community Engagement*

- **Workshops and Events**: Organize workshops, seminars, and events to engage the local community in discussions and activities related to sustainable fashion.
- **Collaboration**: Collaborate with local organizations and schools to promote sustainable fashion initiatives and education.

9.8 *Research and Innovation*

- **Funding Research**: Support research into new materials and technologies that reduce the environmental impact of textiles.
- **Innovation Competitions**: Organize competitions and initiatives to encourage innovative solutions for textile waste reduction.

9.9 *Textile Banks and Collection Points*

- **Establish Collection Points**: Set up collection points where people can drop off old clothing and textiles for recycling.
- **Textile Banks**: Create "textile banks" where individuals can access used clothing for free or at a low cost.

9.10 *Advocacy and Activism*

- **Join or Support NGOs**: Get involved with or support organizations dedicated to reducing textile waste and advocating for sustainable fashion.
- **Petition for Change**: Start or sign petitions calling for fashion industry reforms and better waste management practices.
- Taking action against textile waste is a multifaceted effort that involves consumers, businesses, governments, and communities. We can endeavor to create a fashion industry that is more responsible and sustainable by combining these strategies.

While numerous well-known firms are attempting to reduce textile waste, other organizations and people must also contribute. Here are some ways.

Recycle

When compared to other materials, textiles currently have one of the lowest recycling rates. But these materials—which can include everything from soiled pants and mismatched socks to ripped sweatshirts—can be repurposed into rags, insulation for homes, or even plush toys. By offering in-store recycling bins, retailers like H&M, Madewell, and The North Face streamline textile recycling and make it as easy as going to the mall. Road Runner provides several services for a variety of goods, including contributions of clothing [3].

Organize a Clothing Swap or Exchange

It's common knowledge that "one man's trash is another man's treasure." That same thing applies to your used clothing. Organizing a clothing exchange or swap event is a great way to give your employees new clothes and give them an ethical way to get rid of the items they no longer need [3].

Rent Your Garments

Rent the Runway and similar services save all the resources needed to produce a new garment by offering the option to rent a newly designed outfit for special occasions like job interviews or important business events. By renting, you can also keep up with fashion trends without breaking the bank or having to worry about your clothes piling up in the attic.

Donate

One excellent method to give back and prevent textiles from ending up in landfills is to donate gently used items to charitable organizations. Organizations like ThredUp, The Salvation Army, and Savers will take your used clothing and recycle what they can't sell.

10 Brands Working to Fight Textile Waste

Most people do not ever consider the environmental effects of their clothing. Large quantities of chemicals, water, energy, and other natural resources are needed to produce textiles. The World Resources Institute estimates that 2700 gallons of water are needed to produce one cotton shirt. Throwing away clothing in the trash not only wastes money and resources, but it can take more than 200 years for the materials to break down. Textiles release the greenhouse gas methane during the breakdown stage, and harmful chemicals and dyes seep into the soil and groundwater [13].

A few manufacturers have begun making an effort to prevent garments from ending up in landfills by modifying how their fabrics are produced and by offering consumers better options to dispose of them as a result of environmentally concerned consumers and activists advocating for change in this sector. Following are a selection of the best instances.

10.1 Patagonia

Worn Wear by Patagonia, a digital platform where consumers can purchase, sell in, and sell previously owned Patagonia goods, began in 2017. Patagonia additionally established its initial real pop-up store for Worn Wear in November 2019. The shop supplies a wide range of used items which Patagonia reuses and transforms following obtaining them from its consumers [4]. In the words of Patagonia, "One of the most responsible things we can do as a company is make high-quality stuff that lasts for years, so you don't have to buy greater quantities of it." Patagonia will help to extend the durability of their equipment and prevent additional textiles from ending up in dumpsters by reusing and selling their customers' apparel. Customers may trade their undesired clothing for a shop credit.

10.2 Ecoalf

According to popular belief, Ecoalf uses reused resources in the production of its fabrics, such as coffee grounds, plastic bottles, used fishing nets, and coffee grounds

[12]. Ecoalf's goal is to develop the initial crop of reused products that are equally good as well as desirable as conventional goods. The organization wants to demonstrate how it is not essential to waste the natural assets of our planet. As part of its ongoing efforts to protect the environment, Ecoalf has developed over 300 textiles that are used to make high-end clothing, recovered over 500 tonnes of waste from the ocean floor, and recycled over 120 million plastic bottles.

10.3 H&M

To avoid sending unwanted clothing to landfills, fast fashion giant H&M has installed recycling facilities in more than 4200 of its locations. The first fashion brand to introduce a worldwide clothing line is H&M. The company claims that tonnes of textiles are thrown away annually, even though up to 95% of them have the potential to be reused. Our mission is to make it easy for customers to donate their unwanted clothes to H&M and to support environmental preservation at the same time. Any brand of textile, new or old, is accepted by H&M, which sends it to the closest recycling facility. Customers are given a 15% discount card for each bag of textiles they bring in, which they can use on their next purchase.

10.4 Madewell's

Madewell has an exceptional recycling program. By using them to construct homes, the well-known apparel manufacturer keeps the jeans you wear out of the trash. Cotton's Blue Denim Go Green initiative works with Habitat for Humanity to repurpose donated jeans as insulation for homes. Madewell has prevented the disposal of 415 tonnes of garbage and reused 830,714 pairs of jeans since their partnership started. You can get an American dollar or twenty shop credit from Madewell in exchange for your worn jeans, regardless of the brand [10].

10.5 The North Face

At participating stores, The North Face invites customers to donate their used clothing and shoes to reduce their environmental impact. Because of its clothing, The North Face gathers and sends clothing to Soles4Souls, a nonprofit organization whose goal is to distribute clothing and shoes to those in need while simultaneously creating jobs. As compensation for their gift, customers can apply a $10 reward toward their subsequent purchases. 95,000 kg of clothing and shoes have previously been brought in by customers to be recycled or disposed of.

10.6 Doodlage

India-based Doodlage is the best example of how creativity and environmental awareness can coexist. They make utilization of sustainable resources such as bananas, corn, and organic cotton fabric. Excellent fabric that is left over or discarded from large manufacturing plants is another source of fabric; this is referred to as "wastage" in the trade. Additionally, they discover textiles that, after being cut, several other businesses have decided not to sell. Each of these pieces offers a unique story and fits together like a jigsaw [14].

"With India, Bangladesh, and China accounting for 40% of textile production, these three countries alone generate sufficient waste to make 6 billion garments from scraps and leftovers." We created a brand using cloth that would have otherwise been wasted because of these horrifying statistics.

11 Conclusion

In conclusion, navigating the landscape of integrated solid waste management in the textile industry is not merely a responsibility but a crucial step toward sustainable practices and environmental protection. The intricate web of challenges demands a holistic approach, one that incorporates innovative technologies, strategic collaborations, and a shift in consumer mindset.

Implementing effective waste management practices in the textile sector requires a collective effort from manufacturers, policymakers, and consumers alike. From adopting circular economy principles to investing in eco-friendly materials and processes, the industry must undergo a paradigm shift to mitigate its environmental impact.

Furthermore, a robust regulatory framework plays a pivotal role in shaping the future of sustainable textile waste management. Policies that incentivize recycling, penalize irresponsible disposal, and promote research and development in green technologies can catalyze positive change.

Education and awareness are equally vital components of this transformation. We can build a culture of mindful consumption by educating customers about the adverse environmental effects of rapid fashion and the benefits of making responsible choices. In essence, the journey toward integrated solid waste management in the textile industry is a dynamic process that demands adaptability and commitment. Through collaborative efforts, technological innovation, and informed decision-making, we can weave a fabric of sustainability that not only protects our environment but also paves the way for a greener, healthier future [8].

References

1. Cohen M (2019) Sustainable textile waste management: environmental and social impact of textile waste. Routledge
2. Shaw CB, Carliell CM, Wheatley AD (2002) Anaerobic/aerobic treatment ofcoloured textile effluents using sequencing batch reactors. Water Res 36(8):1993–2001
3. Geng Y, Doberstein B, Fujita T (2012) A review of current knowledge of recycling and reuse of textiles and clothing: a case study in China. J Clean Prod 19(11):1186–1198
4. Khan S, Malik A (2018) Toxicity evaluation of textile effluents and role of native soil bacterium in biodegradation of a textile dye. Environ Sci Pollut Res 25:4446–4458
5. Subramanian N, Gokhale A (2017) Sustainable solid waste management: a systems engineering approach. CRC Press
6. Dhingra R, Forsman A (2018) Circular economy in textiles and apparel: processing, manufacturing, and design. Sustainability 10(3):743
7. Fletcher K (2008) Sustainable fashion and textiles: design journeys. Earthscan
8. Muthu SS, Gardetti MA (eds) (2017) Sustainability in the textile industry. Springer
9. Wang DM (2016) Environmental protection in clothing industry. In: Sustainable development: proceedings of the 2015 international conference on sustainable development (ICSD2015). Columbia University, New York city, USA, pp 729–735
10. Hu QH, Qiao SZ, Haghseresht F, Wilson MA, Lu GQ (2006) Adsorption study for removal of basic red dye using bentonite. Ind Eng Chem Res 45:54–59
11. Kalyani V, Srinivasan K (2013) Study on the impact of textile dye effluent on ground water quality of Erode district, Tamil Nadu, India. Int J Environ Sci 3(2):415–421
12. WHO (2006) Guidelines for the safe use of wastewater. excreta and greywater. Geneva, Switzerland
13. Hassan MM, Carr CM (2018) A critical review on recent advancements of the removal of reactive dyes from dyehouse effluent by ion-exchange adsorbents. Chemosphere 209:201–219
14. Horrocks AR, Miraftab M (2007) Ecotextiles. Woodhead Publishing Limited
15. Das O, Kim NK (2017) Challenges and opportunities in waste management. In: Waste management and valorization. Springer, pp 145–158
16. Dahama AK (2002) Organic farming—an overview for sustainable agriculture (2nd enlarged edition). Agrobios (India), Jodhpur, Rajasthan, India

Role of Governments and Consumers

Sakeena Naikwadi

Abstract The paper discusses the growing awareness of eco-friendliness and sustainable consumption in response to environmental issues. It emphasizes the shared responsibility of individuals, governments, and industries in addressing environmental problems. The government's standards for an eco-friendly environment are highlighted, including the 'Ecomark' scheme and other initiatives. It acknowledges the need for proactive roles from both consumers and manufacturers to implement eco-friendly and clean technology. Despite efforts like the Ecomark scheme, challenges exist due to insufficient response from manufacturers. Government standards, such as green building regulations, renewable energy targets, and waste management regulations, are mentioned as contributors to an eco-friendly environment. It delves into certifications and standards like ISO 9001, GOTS, OCS, etc., within the textile sector, emphasizing their role in advancing sustainability, ethical conduct, and eco-friendly practices. The importance of these certifications in ensuring economic, social, and environmental criteria in the textile industry is emphasized. Additionally, the abstract introduces various certification standards like Green Seal, RWS, RDS, RCS, SFA, and Cradle-to-Cradle, each contributing to sustainable and eco-friendly practices in textile and different sectors.

Keywords Eco-friendly · Environmental issues · Ecomark · Eco-labeling · Ethical practices · Global warming · Government policies · Sustainable consumption · Standards and certifications · Waste management

---

S. Naikwadi (✉)
Department of Textile and Apparel Designing, College of Community Science, University of Agricultural Sciences, Dharwad, India
e-mail: sakeenanaikwadi@gmail.com

Sadhna et al. (eds.), *Climate Action Through Eco-Friendly Textiles*, SDGs and Textiles,
https://doi.org/10.1007/978-981-99-9856-2_11

1 Introduction

Eco-friendly is defined as being earth-friendly and not causing harm to the environment. More people than ever before are realising how directly their actions affect climate change, global warming, and environmental problems. This heightened awareness may be one of the factors driving people to become more conscious of sustainable consumption, the importance of maintaining a positive environment, and the positive effects such actions can have on both society and the environment.

Environmental concerns tend to divide the public into two distinct groups. While those in the second group are deeply concerned about environmental sustainability and actively work to maintain ecological equilibrium, those in the first group tend to hold the government responsible for environmental problems. Even though it may be difficult to accept, society must realize that everyone must work together to address environmental issues [1].

Environmental issues are human-induced [2, 3]. Absolutely, irresponsible human actions that impact the environment demand both behavioral solutions and legislative measures. Since consumers are the ones who drive patterns of consumption, their behavior will have a significant impact on how the future is shaped.

Indeed, sustainability encompasses not only the environmental aspects like pollution, waste, and resource use but also the social aspects including health and welfare, in relation to products developed by various stakeholders such as industries, manufacturers, and artisans. This concept extends its focus to the consumption patterns of both households and governments. It involves the examination of governmental strategies and tools, such as standards, taxes, subsidies, communication campaigns, and education, that are implemented to promote sustainable consumption. Additionally, it addresses methods for safeguarding consumers against misleading information regarding sustainability in areas like labeling, advertising, and corporate reporting, all aimed at promoting sustainable consumption.

Promoting a sustainable environment, consumption, and production stands as pivotal elements in the broader scope of sustainable development. Achieving long-term economic growth that complies with social and environmental standards is essential to sustainable development. Many government policies in this domain primarily concentrate on mitigating the adverse environmental effects stemming from unsustainable industrial production practices, predominantly through the implementation of regulations and taxes. Equally essential is the promotion of sustainable consumption, not only to curtail detrimental environmental and social externalities but also to foster markets for sustainable and eco-friendly products.

The consumption of biosources is rising quickly worldwide as the world economy expands and the global population grows. As a result, environmental issues such as air and water pollution are getting much worse [4]. Worldwide demand for the production and consumption of environmentally friendly and eco-friendly products is rising [5]. As a result, the government encourages and motivates manufacturers to produce green goods [6].

2 Standards by the Government for Eco-friendly Environment

Environmental protection is a topic that has united government, businesses, and consumers on a platform where each must play its part. Legislators and governing bodies are using their power to promote the development of clean technologies and lessen the risks industrialization poses to the environment and public health. But fast industrialization, unplanned urbanization, and shifting consumption patterns brought on by the desire for higher living standards are putting a great deal of strain on the environment. The environment cannot be completely restored to its original state through regulatory actions taken by pollution control agencies alone. Proactive and promotional tactics must be carried out in tandem with an overall environmental protection plan, in addition to these regulatory measures. It is now the responsibility of consumers to lead the way in pressuring manufacturers to implement preventive and mitigating measures as well as clean, green technologies and environmentally friendly ways to dispose of used goods.

2.1 Eco-labeling (Ecomark)

In 1991, the Indian government launched the "Ecomark" eco-labeling scheme to raise consumer awareness. This program's main objective was to make it simpler for customers to recognize products that are environmentally friendly. It required classifying sixteen different consumer product types to create eco-criteria and labeling guidelines. For 14 of these product categories, the government has established formal criteria up to this point. If the producers of these goods satisfy the requirements, they may apply for the eco-label from the Bureau of Indian Standards. The Ministry has also launched awareness campaigns to inform manufacturers and consumers about the program. But the program hasn't taken off because manufacturers haven't responded to it sufficiently.

The scheme's requirements are based on a "cradle-to-grave" methodology, which covers the extraction of raw materials, manufacturing, and disposal. The Ministry has chosen a relatively straightforward system in which the requirements for granting any label should not disregard the quality of the product and should instead concentrate on the direct effects of the product during use and disposal, as well as energy efficiency and noise levels (such as those associated with electrical goods, etc.). The principal environmental standards for products that are listed below are

- Products cause substantially less pollution than other comparable products in production, usage, and disposal
- They are recycled and/or recyclable where comparable products are not
- They make a significant contribution to saving non-renewable resources or minimizing the use of renewable resources compared with other comparable products

- They contribute to a reduction of adverse environmental health consequences
- That their price is not extraordinarily higher than comparable products; and
- They comply with laws, standards, and regulations pertaining to the environment.

To enhance the effectiveness of the Ecomark scheme and encourage wider adoption, the government has also conducted publicity campaigns to raise awareness among consumers and manufacturers. However, it is worth noting that the scheme's popularity has been limited due to insufficient response from manufacturers. The government plays a vital role in setting standards and promoting an eco-friendly environment. Aiming to raise customer awareness, encourage manufacturers to use cleaner technology, and ensure sustainable practices throughout a product's life cycle are initiatives like the Ecomark scheme. By aligning regulatory actions with proactive and promotional approaches, governments can work alongside consumers and industries to achieve a more sustainable future (Table 1).

In addition to the Eco mark scheme, several other government standards and initiatives contribute to promoting an eco-friendly environment:

1. **Green Building Standards**: Governments fixed the building regulations and standards that promote environmental construction techniques, water conservation, the use of sustainable materials, and energy efficiency. By promoting the construction of green buildings, these standards hope to lessen the built environment's negative environmental effects.
2. **Renewable Energy Targets and Incentives**: Governments set targets to raise the share of renewable energy in the overall energy mix. Renewable energy technologies such as solar, wind, hydro, and geothermal energy are encouraged to be used and developed through the provision of incentives such as tax credits, feed-in tariffs, and subsidies.
3. **Waste Management Regulations**: Regulations for waste management are put in place by governments to encourage recycling, waste reduction, and safe disposal of hazardous waste. They set recycling goals, aid in the construction of infrastructure for recycling, and promote the use of recycled materials in production.
4. **Sustainable Transportation Policies**: Governments support alternative fuel development, electric vehicle use incentives, fuel efficiency standards, and investments in public transportation infrastructure to promote sustainable transportation options.
5. **Environmental Impact Assessments (EIAs)**: Environmental impact assessments are a requirement of the government for businesses and construction projects to assess and reduce any potential negative environmental effects. EIAs aid in determining how to reduce pollution, safeguard ecosystems, and promote sustainable development.
6. **Carbon Pricing and Emissions Trading Schemes**: Governments implement carbon pricing tools, like carbon taxes or emissions trading schemes, to incentivize companies and industries to lower their greenhouse gas emissions. Financial incentives are offered by these market-based strategies to encourage the adoption of eco-friendly technologies and the reduction of carbon footprints.

Table 1 Eco-labeling in textile industry

	• **Better cotton initiative**: for cotton production to be more socially, environmentally, and economically sustainable, BCI promotes a comprehensive set of production guidelines
	• **Blue Angel**: certification program promotes both environmental protection and consumer protection by ensuring that products and services bearing the Blue Angel label meet specific environmental and quality standards
	• **BMP certified cotton**: BMP serves as the guiding framework for the Australian cotton industry, fostering the cultivation of cotton in a manner that aligns with our natural environment
	• **Eco mark**: the criteria follow a cradle-to-grave approach, beginning with the sourcing of raw materials and concluding with disposal. Consumer products that meet the defined environmental standards and adhere to the quality benchmarks outlined in Indian Standards are granted the eco mark label
	• **Global organic textile standard**: certified organic fibers must be used in accordance with GOTS. It offers strict environmental and social standards. All processing stages are subject to its criteria
	• **Oeko-Tex Standard 100**: textile products are eligible for Oeko-Tex Standard 100 certification only when all their components meet the required criteria without any exceptions

(continued)

Table 1 (continued)

	• **Green shape**: the Green Shape quality label is exclusively awarded to products that fulfill its criteria, which include being composed of at least 90% organic cotton or recycled materials, dyed using the VAUDE ecolor dyeing process, or manufactured in compliance with the blue sign for textile standards
	• **Eco-INSTITUT**: Eco-Institut supplies clients a reliable and significant label for building products and textiles without any health hazards

7. **Conservation and Protected Area Management**: To preserve biodiversity, safeguard ecosystems, and encourage sustainable land use practices, governments create and manage protected areas, national parks, and wildlife sanctuaries. To stop habitat destruction, illegal wildlife trade, and unsustainable resource extraction, they pass laws and regulations.
8. **Research and Development Funding**: Governments allot money for research and development into environmentally friendly technologies, sustainable farming methods, reducing climate change, and protecting the environment. This funding promotes innovation, scientific development, and the creation of remedies for environmental problems.
9. **International Collaboration and Agreements**: To collectively address international environmental issues, governments take part in international partnerships and agreements. To reduce greenhouse gas emissions, safeguard biodiversity, and advance sustainable development on a global scale, they participate in negotiations and make commitments.

    By implementing these government standards and initiatives, policymakers aim to create a regulatory framework that encourages businesses, industries, and individuals to adopt eco-friendly practices, reduce environmental impact, and contribute to the long-term sustainability of the planet.

3 Certifications and Standards of the Textile Sector

Every industry strives for success within its specific domain. However, true success can only be attained when three fundamental criteria are met: economic, social, and environmental. In the textile industry, certifications are founded upon these assessments. Certification serves as a guarantee that rigorous standards are upheld, broadens the industry's customer base, and evaluates the effectiveness of the quality management system.

Textile testing is a vital process employed worldwide to assess various textile materials following specific requirements. It serves as a means for companies, consumer organizations, and governments to ensure the safety, quality, and affordability of textile materials. Within the textile industry, certifications and standards play a pivotal role in advancing sustainability, ethical conduct, and environmentally conscious production methods. These certifications empower consumers to make well-informed decisions, establish transparency within the supply chain, and incentivize textile manufacturers to embrace responsible and eco-friendly practices. Here are some well-known certifications and standards within the textile sector.

3.1 International Organization for Standardization (ISO) 9001 (2015)

ISO 9001 focuses on establishing a robust quality management system, a critical element for organizations of all sizes, irrespective of their industry. ISO-certified companies can be found in over 170 countries globally. This standard is built upon several key quality management tenets, such as a dedicated customer focus, top management involvement, a process-oriented methodology, and a continuous improvement mindset.

ISO certification is viewed as a crucial prerequisite for some textile and apparel businesses before they can engage in international export activities. Any organization bearing an ISO certification signifies its commitment to maintaining consistency and delivering high-quality products and services, all of which collectively contribute to building trust with customers and stakeholders [7].

3.2 Global Organic Textile Standard (GOTS)

For the processing of textiles made from organic fibers, the Global Organic Textile Standard (GOTS) is an internationally recognized standard. It encompasses both ecological and social criteria and ensures the traceability and integrity of organic textiles throughout the entire supply chain. Here are key aspects of the GOTS standard:

1. Organic Fiber Criteria: A minimum percentage of organic fibers must be present in a textile product to comply with GOTS requirements. The standard outlines the requirements for organic farming, including the use of approved techniques and materials to cultivate fibers without the use of synthetic pesticides, fertilizers, or pesticides, as well as genetically modified organisms (GMOs).

2. Environmental Criteria: Environmentally friendly production and manufacturing techniques are promoted by GOTS. It places strict restrictions on the use of harmful substances, such as prohibitions on formaldehyde, heavy metals, and aromatic solvents. Energy-saving measures and waste-water treatment systems are also mandated by GOTS.
3. Social Criteria: To guarantee a fair wage and good working conditions, GOTS includes social criteria. Forced labor, child labor, and discrimination are all forbidden. The standard preserves employees' rights to collective bargaining and freedom of association while promoting just compensation and safe working conditions.
4. Supply Chain Traceability: According to GOTS, all organic fibers must be fully traceable from their point of origin to the finished product. It makes sure that the textiles are processed, manufactured, and labeled under organic standards and are approved by recognized certification bodies.
5. Certification and Labeling: Independent third-party certification organizations handle GOTS certification. Manufacturers and processors of textiles can get GOTS certification for their goods, which enables them to export organic fabrics and apparel with a single certification that is recognized all over the world. Products with the GOTS certification can be identified as such on the label, giving customers confidence and transparency.

The GOTS allows consumers to buy textiles with confidence, knowing that they are produced sustainably and with social and environmental welfare in mind. It helps the organic textile industry grow and enables consumers to make knowledgeable decisions. By adhering to the GOTS standard, textile manufacturers and processors support ethical and sustainable business practices in the textile industry and protect the integrity of organic textiles along the entire supply chain.

3.3 Fair Trade

Promoting just and durable trade relations between consumers in developed countries and producers in developing countries is the primary objective of the Fair Trade certification and social movement. By ensuring that marginalized and disadvantaged producers—often those employed in sectors like agriculture and handicrafts—are fairly compensated for both their labor and their goods, it seeks to establish a more just and equitable global trading system. To enhance working conditions and safeguard the environment during the production process, Fair Trade also advocates for social and environmental standards. A global network of organizations that support and advocate for fair trade practices makes up the World Fair Trade Organization (WFTO). Here are key aspects of Fair Trade certification:

1. Objectives: By guaranteeing fair prices, secure working conditions, and community development, fair trade seeks to open up opportunities for marginalized producers in developing nations, particularly in the agricultural sector. It aims to

support sustainable livelihoods, reduce poverty, and strengthen the position of producers.

2. Product Scope: Textiles are among the many final goods that can be certified as Fair Trade. Fair Trade cotton, which is produced by farmers who receive fair prices for their crops and work in decent and safe conditions, is frequently used in the production of fair trade textiles.
3. Supply Chain requirement: Fair Trade certification places a strong emphasis on transparency and traceability throughout the entire supply chain. Individuals and entities at various stages, including producers, grower groups, processors, importers, exporters, brands, and distributors, have the option to seek Fair Trade certification based on their specific roles in the supply chain. This certification serves as a guarantee that fair trade principles are adhered to at every step of the production, processing, and distribution process.
4. Fair Pricing: By guaranteeing a livable wage and paying producers fairly for their goods, fair trade helps to offset the costs of producing goods in an environmentally responsible manner. This ensures that producers can invest in their companies, communities, and future development while also addressing the imbalances in conventional trade.
5. Social and Environmental Standards: Social and environmental standards are included in fair trade certification. It guarantees that employees have the right to fair compensation, flexible scheduling, and the freedom to associate. It also ensures that employees are given safe and healthy working conditions. With recommendations to reduce the use of dangerous chemicals, safeguard ecosystems, and support sustainable farming practices, environmental sustainability is also encouraged.
6. Community Development: The importance of community empowerment and development is emphasized by fair trade. Producers and their communities participate in decision-making and gain from the Fair Trade premium, which is used to fund local initiatives like infrastructure, healthcare, and education.
7. Certification and Labeling: Independent certifying organizations handle the certification of fair trade products. Once certified, products can display the Fair Trade label, giving consumers peace of mind that the item satisfies Fair Trade requirements. Customers can choose ethically when they shop thanks to the label, which also helps producers in underdeveloped nations.

By choosing Fair Trade-certified textiles, consumers can contribute to sustainable and ethical practices in the textile industry. Fair Trade certification ensures that the products they purchase have been produced under fair conditions, benefiting producers and their communities.

3.4 ECO PASSPORT by OEKO-TEX

ECO PASSPORT by OEKO-TEX is an essential independent testing and certification program created especially for the chemicals, colorants, and other additives used in the textile industry. This system confirms that these chemical products and the components that make them up adhere to specific standards for safety, sustainability, and legal compliance. The certification process consists of three stages:

1. Checking chemical products and ingredients against the Manufacturing Restricted Substance List (MRSL) and OEKO-TEX Restricted Substance List (RSL) is the first step. These lists include substances that, due to possible risks to human health and the environment, are either prohibited or subject to restrictions.
2. Analytical Verification: In the second stage, analytical tests must be carried out at a lab that is a member institute of OEKO-TEX. Through this verification process, it is ensured that the chemical ingredients and products satisfy safety, sustainability, and regulatory compliance requirements. The tests quantify the amount and concentration of potentially hazardous substances.
3. Verification of Product Stewardship Measures: The third step evaluates the product stewardship procedures used by the manufacturer. This also means assessing their working conditions and environmental management strategies to guarantee moral and sustainable production practices.

A list of textile products that successfully fulfill the requirements for the ECO PASSPORT by OEKO-TEX certification can be found in the OEKO-TEX Buying Guide. The Buying Guide acts as a central sourcing platform for products and materials that have already been approved. It offers a useful tool for manufacturers and buyers to locate textiles that have been approved as meeting the exacting requirements of OEKO-TEX's ECO PASSPORT.

Consumers and the textile industry can feel reassured by OEKO-TEX's ECO PASSPORT certification that the chemicals used in textile production have undergone safety, sustainability, and regulatory compliance testing and certification. It advances the overarching objective of encouraging socially and environmentally conscious practices throughout the textile supply chain [8].

3.5 SA8000

The SA8000 Standard is the world's greatest social certification program. Upholding the highest social standards, the SA8000 Standard and Certification System guarantees that workers receive fair and decent treatment. SA8000 assesses several important aspects, such as freedom of association, health and safety, child labor, and collective bargaining rights.

3.6 *Worldwide Responsible Apparel Production (WRAP)*

Following the establishment of the first Board of Directors in 1999, WRAP was formally incorporated in 2000. WRAP is designed with the express intent of remaining independent of the clothing industry in terms of governance and finances. Its primary goal is to use certification and educational initiatives to advance safe, legal, humane, and ethical manufacturing practices across the globe. The WRAP certificate is a widely accepted indication of a commitment to upholding moral and social norms.

3.7 *Blue Sign*

The "blue sign" standard signifies a commitment to resource management that minimizes adverse impacts on both people and the environment. This certification standard encompasses aspects such as consumer safety, prevention of water and air pollution, occupational health, and the responsible use of hazardous substances.

Manufacturers and brands that adhere to the blue sign standard are obliged to operate responsibly and sustainably, considering the welfare of individuals, the environment, and resource conservation. The blue sign indicates that the textile product was produced in an environmentally sustainable manner and that it satisfies strict international consumer safety regulations. In addition to lowering human risk, the widespread use of chemicals certified by the Blue Sign program also saves money.

3.8 *Zero Discharge of Hazardous Chemicals (ZDHC)*

A cooperative effort, the Zero Discharge of Hazardous Chemicals (ZDHC) program seeks to end the release of dangerous chemicals along the textile, leather, and footwear value chain. It aims to safeguard consumers' and employees' health and well-being while enhancing environmental sustainability. Here are key aspects of the ZDHC program:

1. Objectives: The main goal of ZDHC is to stop hazardous chemical discharge in the textile, leather, and footwear industries at every stage of production. Toxic materials must be removed from the environment and dealt with at the source to achieve this.
2. Contributor Categories: Three contributor categories make up the ZDHC program: Signatory Brands, Value Chain Affiliates, and Associates. Companies that agree to follow the objectives and principles of the ZDHC program are known as Signatory Brands. Suppliers, chemical suppliers, and solution providers that actively support and take part in ZDHC activities are referred to as value chain affiliates. Associate organizations support the goals and tenets of the program.

3. Multi-Stakeholder Approach: ZDHC is a multi-stakeholder organization that works with over 170 contributors from different industry sectors. Brands, distributors, chemical suppliers, and service providers fall under this category. ZDHC seeks to promote cooperation and information exchange to establish best practices and spur industry-wide advancements.
4. Chemical Management: ZDHC concentrates on encouraging ethical chemical management techniques. This entails creating and putting into practice standards, tools, and guidelines to assist businesses in identifying, replacing, and removing hazardous chemicals from their production processes. To decrease the effects of hazardous chemicals in the atmosphere and human health, it also emphasizes proper handling, storage, and disposal of chemicals.
5. Quality Assurance: To increase consumer well-being, ZDHC advocates for appropriate quality control at every stage of production. ZDHC strives to guarantee that footwear, leather goods, and textiles satisfy safety and legal requirements so that consumers can feel confident about the products they purchase. By addressing the existence of dangerous chemicals in products, this is accomplished.

Through the partnership, ZDHC hopes to create a textile, leather, and footwear sector that is more ethical and sustainable. Through encouraging the adoption of best practices, fostering transparency, and providing guidance and resources, ZDHC seeks to mitigate the detrimental effects that hazardous chemicals have on the environment and human health along the value chain.

3.9 Responsible Care

The chemical industry's voluntary Responsible Care program aims to increase public awareness of environmental, health, safety, and security-related issues. It also emphasizes enhancing the lifetime efficiency of technologies, stages, and products to prevent harm to people and the environment. Responsible Care engages with stakeholders to better understand and address their needs in addition to keeping an eye on things.

3.10 Registration, Evaluation, Authorization, and Restriction of Chemicals (REACH)

Encouraging better environmental and public health protection is the aim of the Registration, Evaluation, Authorization, and Restriction of Chemicals (REACH) regulation. This is accomplished by making it easier to identify the intrinsic properties of chemical substances at an earlier age and with greater accuracy. REACH also aims to increase innovation and competitiveness in the chemicals sector of the European Union.

One of its main goals is to reduce the use of dangerous materials by encouraging the replacement of dangerous chemicals that are frequently used in both daily products and industrial processes. Ensuring that accurate information about chemicals used in industries is easily accessible is one of REACH's main goals, especially when those substances may be harmful to human health.

3.11 Green Seal

The goal of the global non-profit organization Green Seal is to promote a more ecologically sustainable and healthy world. It promotes sustainability and upholds strict standards for product performance, environmental sustainability, and human health. Green Seal, which was founded in 1989, has led the way in the movement for environmental labeling.

The 2014 ISEAL Standard-Setting Code states that Green Seal's standard-setting process complies with internationally recognized best practices. This commitment to the code guarantees transparency throughout the entire production process.

3.12 Organic Content Standard (OCS)

The OCS (Organic Content Standard) certification is a widely accepted standard that attests to the use of organic material throughout a product's processing stage. It guarantees the existence and proportion of organic material in the finished product. Here are some key points about the OCS certification:

1. Organic Material Verification: The product's specific percentage of certified organic material is guaranteed by the OCS certification. The product must contain at least 95% certified organic fiber to meet the OCS 100 standard. Except for leather products, this requirement applies to all fibers, wool, fabrics, apparel, and upholstery fabrics.
2. Processing Stage Focus: The OCS focuses on confirming the utilization of organic material during the product's processing stage. It guarantees that the product's organic fibers, from raw material sourcing to manufacturing, have been processed under organic processing standards.
3. Certification and Labeling: A product may display the OCS label, which indicates the presence and percentage of organic material, once it has obtained OCS certification. This label enables customers to recognize and select products with the organic content they prefer with knowledge.
4. Organic Fiber Transparency: The OCS certification encourages supply chain transparency by monitoring and confirming the presence of organic fibers. It assures customers that a product's organic claims are supported by third-party certification.

5. Consumer Confidence: Customers can be confident that the product they are buying has a high percentage of organic material thanks to OCS certification. This certification encourages sustainable and ethical practices in the textile industry and supports consumer demand for organic products.

The OCS certification helps consumers make educated decisions and promotes the expansion of the organic textile industry by attesting to the percentage of organic material in a finished article. It contributes to a more sustainable and environmentally conscious textile industry by giving transparency and credibility to claims of organic content.

3.13 FLOCERT

An international certification and verification agency with a focus on Fair Trade goods is called FLOCERT. It makes it possible for businesses to follow moral business principles and guarantee fairness in their supply chains. A company's visibility and reputation on the international market are improved by obtaining FLOCERT certification, which demonstrates the company's dedication to ethical and fair trade principles.

3.14 Responsible Wool Standard (RWS)

Farmers have the opportunity to produce wool of the highest quality while upholding international standards thanks to the Responsible Wool Standard (RWS). RWS places a high priority on protecting the quality of the grazing land that sheep use as well as their humane treatment.

Having an RWS certificate is beneficial because it attests to the moral standards followed at the farm level. With the help of this certification, brands now have a clear way to make claims about where their wool comes from.

3.15 Responsible Down Standard (RDS)

In the textile industry, exploitation is a common worry, particularly when sources derived from animals are involved. The Responsible Down Standard (RDS), which ensures animal welfare in products containing down and feathers, is put in place to address this issue. This standard ensures that animal welfare is given top priority in the textile industry and applies not only to down and feathers but also to other animal products like wool, angora, cashmere, and leather.

3.16 Recycled Claim Standard (RCS 100)

It aids in ensuring that recycling procedures are optimized to reduce production-related waste. RCS enhances transparency and ethical recycling practices within the sector by verifying the presence and quantity of recycled materials in the end product.

3.17 Sustainable Fibre Alliance (SFA)

The Sustainable Fibre Standard (SFA) was created to address sustainability issues in cashmere production. It places a strong emphasis on ensuring the well-being and safety of goats and herders involved in the process. Introduced in 2015, it stands as the world's pioneering comprehensive sustainability standard for cashmere.

The SFA standard draws its foundation from the Content Claim Standard developed by Textile Exchange, a standard that supports numerous other sustainability certifications and has already gained industry-wide adoption.

3.18 Cradle to Cradle Certification

The multi-attribute label you're describing provides a mechanism to showcase eco-intelligent product design efforts. It places particular emphasis on aspects such as the use of environmentally friendly materials, recyclability, and efficient utilization of water and energy, among other factors. This label applies not only to finished products but extends from raw materials to the end product, ensuring a comprehensive approach to sustainability and eco-conscious design.

4 Conclusion

To make an eco-friendly environment and to bring sustainability, manufacturers, industrialists, artisans, etc., should follow the Government rules and regulations. A sustainable corporation needs to convert raw materials into products that people appreciate for their positive impact on quality of life and their role in preserving the environment. The environmental label is poised to become a significant component of future environmental policies and will demonstrate its value as an effective tool, rooted in cooperation and voluntary commitment from the industry. The certifications and standards help consumers identify textiles and garments that have been produced with sustainability and ethical considerations in mind. They encourage textile manufacturers to adopt responsible practices, reduce environmental impacts, and improve worker welfare. By seeking out products with these certifications, consumers can contribute to a more sustainable and socially responsible textile industry. These certifications and standards developed by the Government of India would protect the environment from hazardous pollution and make the green environment.

References

1. Yahya WK, Musal ND, Hashim NH (2017) Understanding environmental friendly consumer behavior. https://www.researchgate.net/publication/305726947
2. Oskamp C, Saunders C (2003) The emerging field of conservation psychology. Hum Ecol Rev 10:137–149
3. Ramly Z, Hashim H, Yahya WK, Mohamed SA (2012) Environmentally conscious behavior among malaysian consumers: an empirical analysis. J Pengurusan 35:111–121
4. Meng Q, Li M, Li Z, Zhu J (2020) How different government subsidy objects impact on green supply chain decision considering consumer group complexity. Math Probl Eng 1–12
5. Dangelico RM (2017) What drives green product development and how do different antecedents affect market performance? A survey of Italian companies with eco-labels. Bus Strategy Environ 26(8):1144–1161
6. Yuan P, Dong Z, Xu J, Lin X (2021) How government regulations and consumer behavior influence manufacturers product green degree decision-making: an agent-based model. Hindawi Wirel Commun Mob Comput 1–18. https://doi.org/10.1155/2021/5582140
7. Banerjee (2019) https://textilefocus.com/brief-certifications-required-textile-industry/
8. Piya FA (2019) A brief on certifications required in textile industry. https://textilefocus.com/brief-certifications-required-textile-industry

Eco Textiles: The Present and the Future

Manpreet Kaur, M. Pavan, and Lata Samant

Abstract Industrial and financial development plays an important role in any nation's progress. When it comes to the implementation of new sustainable practices, the textile sector is currently undergoing a period of evolution means making steady and gradual progress toward adopting more ecologically oriented and socially sensitive practices. The purpose of this paper is to examine how sustainable innovations and wastewater management are currently being applied in the textile industry. Textiles are an essential part of our everyday lives and are significant to the world economy. It is commonly acknowledged that the textile industry is among the most difficult to integrate sustainability into its operational framework. There are several implementations such as innovation in sustainable products including sustainable production and packaging materials, life cycle assessment, Eco-design, and Eco-labeling. Textile sustainable process innovation includes the use of enzymes in textile processing, improved waste management techniques, eco-efficiency, supply chain management, and sustainable textile product production. The acquisition of environmental management systems, corporate policies, cooperative practices between businesses and consumers, distinctive business models, culture, and knowledge systems are additional examples of organizational sustainability innovation.

Keywords Textile industry · Product · Process and organizational innovations · Textile Sustainable innovations

M. Kaur (✉)
The Maharaja Sayajirao University of Baroda, Vadodara, Gujarat 390002, India
e-mail: mannukhinda08@gmail.com

M. Pavan
Punjab Agricultural University, Ludhiana, Punjab 141004, India

L. Samant
G. B. Pant University of Agriculture and Technology, Pantnagar, Uttarakhand 263145, India

Sadhna et al. (eds.), *Climate Action Through Eco-Friendly Textiles*, SDGs and Textiles,
https://doi.org/10.1007/978-981-99-9856-2_12

1 Introduction

Swachh Bharat Abhiyan movement rapidly gained movement throughout India, a great step toward a cleaner and sustainable future for the nation. To improve environmental and social well-being, this nationwide campaign was initiated to promote cleanliness and hygiene addressing the pressing issues of waste management and population. As per India's commitment to the Paris record, the nation is currently prepared to take the next significant step toward a sustainable future. This new stage aims to manage trash and pollution in a way that goes beyond the traditional methods. Rather, it sees these obstacles as possible sources of energy and growth opportunities for the economy. This novel approach not only tackles environmental issues but also maximizes their potential to support the sustainable growth of the country.

The United Nations Agenda 2030, along with its set of 17 Sustainable Development Goals (SDGs), global conversion on sustainable development has gained an additional boost. It has completely changed the concept of sustainability [1]. This concept defines sustainable development as an ongoing process that entails a more conscious and effective use of natural liabilities.

Sustainability is defined as meeting present consumer needs without affecting those of future generations. The global carbon dioxide emissions from the textile business are enormous, overtaking those from several other industries like international shipping and aviation. In recent decades, the world's population has increased significantly, and during the same time, general living standards have also increased. Due to these two advancements, there is now more demand for textiles, which has raised the level of textile production. One of the people's basic requirements in the present society is textiles.

The textile and apparel industry is regarded as one of the earlier, several hundred-year-old segments of the Indian economy that encountered the formidable obstacles of implementing sustainable business practices. Indian textile sector is different from other industries due to the close relationship between the textile and agriculture sectors. The sustainability of the textile industry needs to be taken into consideration by keeping in mind the current global climate change and the growth or development of the population. Second to the oil industry, the textile industry is considered a polluting sector in the world because it pollutes our planet at every stage of its production. It emits large amounts of greenhouse gases (approximately 1.2 billion tonnes), compared to all international travel and maritime shipping, but it also contributes to the economy of India as it contributes to exports, industrial production, foreign exchange, and employment. The industry had a positive effect on the economy, while resource depletion, pollution, and greenhouse gas emissions harmed the environment [2].

The usage of harmful chemicals, high energy, and water consumption, significant waste production, labor-intensive transportation, and an abundance of packing materials are responsible for the unsustainability of the life cycle of textiles and apparel. Up to 25% of global carbon emissions are expected to come from the fashion sector by 2050. The fashion and apparel industry's detrimental effects on sustainability necessitate the implementation of production and consumption strategies and practices.

The textile industry is criticized for causing harmful environmental effects because of its supply chain operations (including the generation of waste, consumption of resources, and the production of a carbon footprint) [3]. The production of textiles and apparel requires a considerable amount of energy, water, and eco resources, which contributes to the rapid increment of textile waste. In addition, it is anticipated that the quantity of waste produced globally from textiles will increase by sixty percent each year between the years 2015 and 2030, leading to a total of 148 million tonnes of trash being produced [3, 4]. In this sense, the depicted representation in Fig. 1 makes more appropriate in terms of circular industry. Some of the possible implementations pertaining to product process and organization innovation are depicted in Fig. 2.

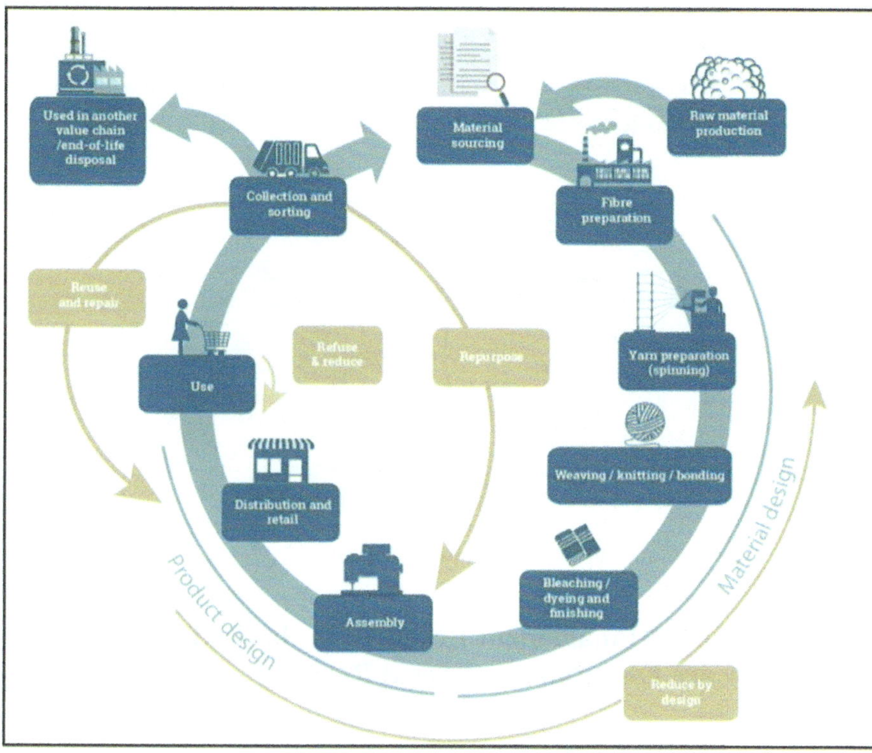

Fig. 1 Textile value chain development of sustainability and circularity UNEP [5]

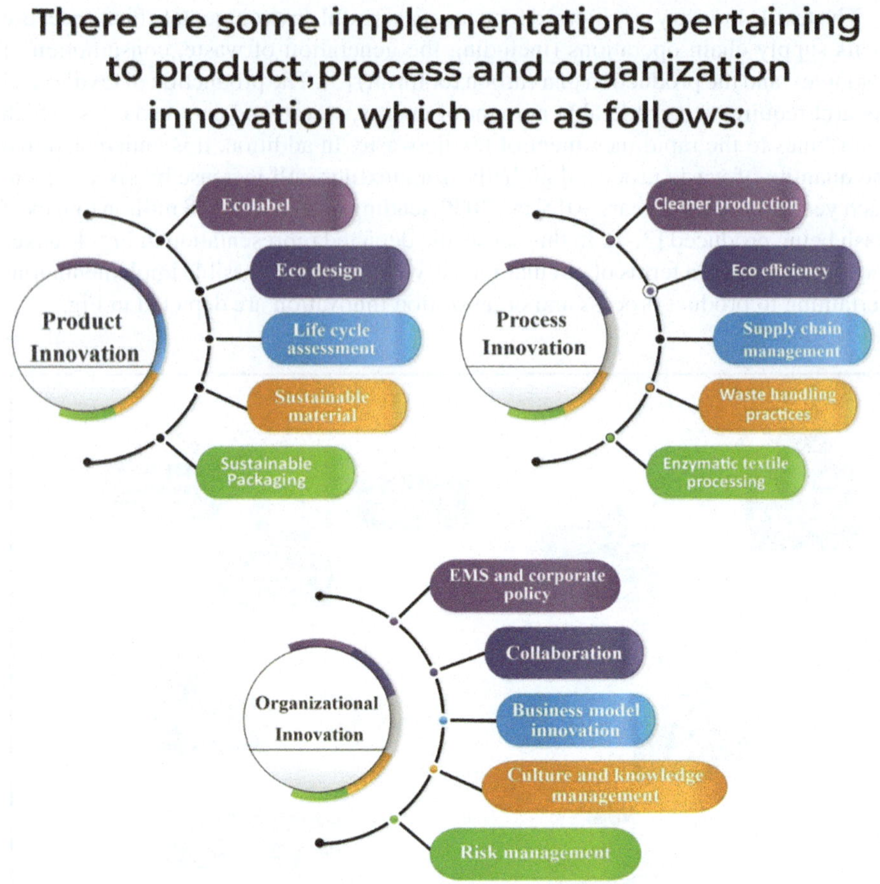

Fig. 2 Sustainable innovation in textile and apparel industry

2 Better Waste Management Practices in the Textile Industry are Essential for Several Reasons

2.1 Environmental Conservation

The apparel industry is recognized among the most environmentally damaging sectors globally may be due to excessive consumption of water, energy, and raw materials which not only depletes natural resources but also results in several environmental problems. It consumes an enormous quantity of water, mainly in water-scarce regions, and relies heavily on energy and raw materials. In the meantime, water pollution and carbon emissions are caused by the textile industry's production methods, and use of dangerous chemicals, and dyes. These pollutants can harm ecosystems and damage natural resources including bodies of water and soil. Additionally textile industry generates a substantial amount of waste both in gaseous and solid form including discarded fabric scraps, chemical byproducts, and emissions from production processes. The transportation of textiles, often to distant locations where textile units are situated, requires a significant amount of fuel. This contributes to carbon emissions and air pollution, especially when long-distance shipping is involved and the amount of high fuel consumption for transportation to long-distance places where textile units are located.

2.2 Resource Conservation

Excessive amounts of water and other natural resources, along with raw materials like cotton and polyester, are needed for the manufacture of textiles. It is possible to recycle and reuse these resources, so lowering the need for new ones and fostering sustainability, by putting improved waste management follows into place.

2.3 Reduction of Landfill Wastes

The textile industry generates a substantial amount of trash, encompassing debris, defective products, and fabrics that have reached their usefulness. The removal of these materials from landfills can be achieved by the utilization of recycling and waste-to-energy technologies, eventually decreasing the pressure on waste management practices. Another change in disposal practices is the throwing away of unneeded clothing rather than donating it.

2.4 Energy Efficiency

By using various techniques like recycling and waste-to-energy conversion, which use less energy than raw material extraction and processing, proper waste management, can result in energy savings. This helps the industry become more environmentally friendly and energy-efficient.

2.5 Regular Compliances

Governments and international bodies are increasingly imposing strict regulations on waste management and environmental practices in the textile industry. Complying with these regulations is essential for avoiding fines, legal issues, and reputational damage.

3 Product Innovation

3.1 Eco Design

Ecodesign is an organized way to integrate environmental considerations with the production of products and processes. Its primary purpose is to present shreds of proof, to verify that a garment is ecologically sound or not, and eventually provide help to the consumer in making an informed and right decision about their purchase [6]. Comparing standard protocol of product designing, ecodesign takes a more in-depth because it executes the whole supply chain (from the raw material up to the consumption of the product) while taking into consideration how the design serves a role for durability and recycling and also considers the recycling element, how the design itself may be recycled [7].

For a producer to receive the "Green" certification for their product, they must fulfill the requirements of an eco-label, which promotes the environmentally responsible production and usage of all resources.

Through a review of three prominent international fashion and apparel companies, it was determined that design plays a significant role as an environmental competency within the sector. This competency encompasses the Five R's, which are.

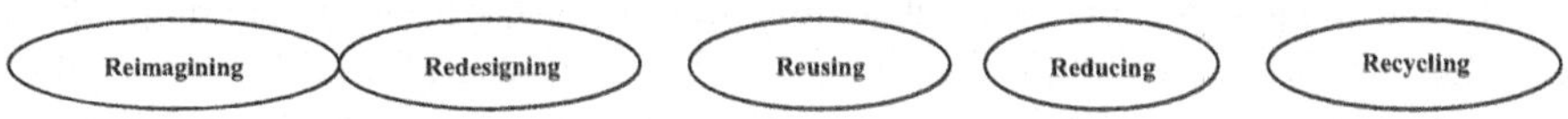

3.2 Environmental Quality Competition and Eco-labeling

In recent years there has been a noticeable shift in the industrial sector particularly within the textile industry toward embracing more environmentally responsible manufacturing processes. This shift is driven by growing awareness of the need for sustainable production and consumption. A significant advancement in this shift is the extensive implementation of eco-labels. Eco-labels have become a prominent element that offers consumers and businesses direct evidence regarding the environmental sustainability of a product. The aforementioned labels serve as a prominent indicator of a product's adherence to particular environmental standards and procedures, providing transparency and reassurance to individuals who give priority to the environmental influence of their consumer choices. Thus, eco-labels serve severe essential functions.

3.2.1 Consumer Information

Eco-labels provide consumers with important details regarding the environmental characteristics of a given product. Consumers can enhance their decision-making process by actively looking out and considering labels that assist in the identification of products that are in line with their environmental principles.

3.2.2 Environmental Certification

It is a process whereby products are assessed and assigned eco-labels based on their adherence to specific sustainability criteria and eco-friendly production standards. These certifications demonstrate that the manufacturer has taken steps to reduce the product's environmental footprint.

3.2.3 Market Differentiation

Eco-labels can provide enterprises with a competitive advantage. Products that have eco-labels have the potential to differentiate themselves in the market by engaging environmentally conscious consumers, thereby potentially acquiring a competitive advantage.

3.2.4 Encouraging Sustainable Practices

The existence of eco-labels serves as an incentive for enterprises to adopt sustainable and environmentally friendly industrial practices. To acquire and sustain these labels, producers must consistently adhere to and exceed predetermined environmental standards.

3.2.5 Environmental Accountability

The concept of environmental responsibilities can be seen through the use of eco-labels, which serve as a means to ensure that manufacturers stay responsible for their promises of the eco-friendliness of their products. The creation of accountability mechanisms promotes trust among consumers and ensures that products effectively contribute to environmental preservation.

As per the Global Eco-labeling network, a product receives an eco-label (a mark that is legally protected) when a product meets a predetermined set of environmental and social standards. The eco-label denotes the product's overall environmental impact which is selected by considering its life-cycle assessment, thus serving to acknowledge the product's environmental leadership.

Eco-labels can be classified into different categories. Some eco-labels take into account the environment and social welfare, whereas others only take into account the environment or social welfare [8]. Ecologically sound practices, environmental safety, product recycling, biodegradability, ozone friendliness, and low energy consumption are all associated with eco-labels [9].

Eco-labels promote environmentally friendly practices for sustainable production and decrease the usage of harmful chemicals and dyes in textiles. The textile sector commonly adheres to a recognized eco-label that requires certified enterprises to use sustainable raw materials, production processes, and technology, reduce water consumption, and less textile effluent water discharge [10].

3.3 Eco-label Objectives

3.3.1 To Avoid Misleading Environmental Advertising

Eco-labels such as Bluesign are designed to address the issue of fraudulent environmental promises. Historically, several corporations have been involved in the practice of "greenwashing," wherein they have exaggerated or spread misleading information about their products, claiming them to be environmentally sustainable. Eco-labels offer a standardized and trustworthy mechanism for customers to identify products that conform to environmental criteria.

3.3.2 To Educate and Motivate Consumers for Green Product Practices

Eco-labels serve as a key mechanism for increasing consumer consciousness regarding environmentally friendly choices. When consumers observe the Bluesign label on a textile product, they can infer that the item has undergone a comprehensive examination to determine its environmental implications. The aforementioned consciousness enables consumers to make more informed decisions and judgments that are better informed and opt for those products that are in alignment with their environmental principles.

3.3.3 To Give Market-Driven Incentives with Less Adverse Environmental Impacts on Products and Manufacturing Processes

Eco-labels give businesses incentives based on the market. Through the acquisition of these positions, corporations can establish product differentiation and gain a competitive advantage. Textile manufacturers that adhere to the Bluesign guidelines have the potential to attract consumers that prioritize sustainability eventually increasing sales and market share. The phenomenon generates a favorable financial incentive for enterprises to embrace more environmentally friendly practices.

The clothing and apparel industry includes many eco-labeling systems that aim to promote environmentally responsible practices. These schemes facilitate the identification of products that comply with certain ecological standards for both manufacturers and consumers. Although there is no one universally recognized eco-label for a particular clothing product, the textile industry offers numerous eco-labels that are specifically designed to address various problems within this sector.

The Eco-label Index is a globally recognized and comprehensive database that offers a complete overview of the different eco-labels utilized in various industries. At present, there is a compilation of 455 eco-labels emanating from approximately 199 countries, ranging across 25 different industry sectors. Therefore, the textile sector is supported by a total of 107 eco-labels, which cater to both global and regional markets. An overview of the global eco-label based on application and common environmental criteria assessed is given in Fig. 3.

The categorization of eco-labels by the International Standard Organization (ISO) encompasses three broad classifications, denoted as **Type I, Type II, and Type III**. These categories serve to clarify the characteristics and extent of eco-labels.

Type I eco-labels refer to a specific category of environmental labels that are used to identify and certify products or services that have met certain predefined environmental criteria.

The organization adheres to rigid standards for environmental quality as defined by the ISO 14024 standard. It also evaluates the environmental consequences encompassing the complete life cycle of a particular product. This includes various elements, i.e. energy use, utilization of water, textile waste management, and carbon footprint.

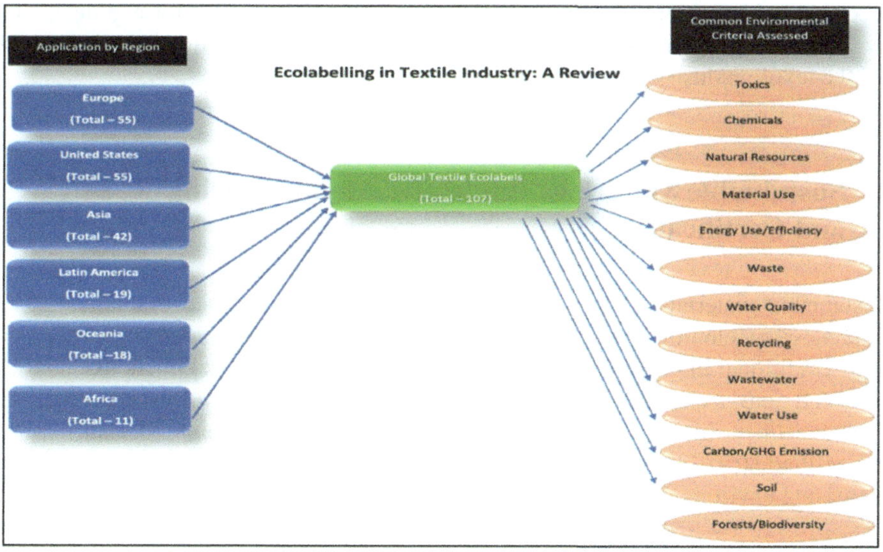

Fig. 3 Eco-labeling in the textile industry [11]

Fig. 4 Eco-labels used in the textile industry [12]

Type I eco-label: thorough assessment of a product's environmental impact, tracing it back to the raw material extraction and preparation process and ending with the consumer, buyer, or user's final consumption.

Type II eco-label: a particular class of environmental labels intended to give customers accurate and thorough information regarding the benefits and drawbacks of a product or service on the environment. It also applies to institutional designations that are not bestowed by a separate governing authority. On the other hand, direct releases by producers or manufacturers are typical.

The aforementioned labels are generally characterized by a lower level of discipline compared to Type I labels, as they tend to concentrate on particular traits or statements, such as "biodegradable" or "organic." Nevertheless, it is possible that they may not possess the third-party verification that is common to Type I and Type III labeling.

Type III eco-labels: a specific classification system that is used to execute and verify the environmental friendliness of products, services, and organizations. Type III eco-labels encompass a comprehensive evaluation of distinct environmental characteristics, which are carefully verified by an independent third-party authority. These sources offer comprehensive and substantiated data regarding the ecological characteristics and impact of a given product. The distribution of this data is frequently facilitated by the utilization of Environmental Product Declarations (EPDs) or similar records, enabling both consumers and businesses to make well-informed decisions by considering extensive environmental information. Some of the eco-labels used for the textile industry are mentioned in Fig. 4.

4 Sustainable Materials, Innovative Manufacturing and Certification

Sustainable materials used in the textile industry have changed dramatically, as awareness of environmental and social issues has increased among the population, Traditional Indian textiles were primarily made from natural fibers like cotton, silk, and jute which were biodegradable and eco-friendly. However the use of synthetic fibers like polyester and nylon has become more common contributing to non-biodegradable waste that takes a long time to decompose which results in an increasing environmental population. Because textile materials used in the industry, such as fibers derived from synthetic and fossil fuel sources, are highly unsustainable, sustainability innovation in the textile industry is crucial [13]. The concept of sustainability is being emphasized by the textile fashion industry more and more; naturally sustainable textile fibers with no environmental impact are being promoted. Although organic cotton and Wool have been used for many years, their acceptance as a sustainable option has grown just recently, in recent times they have been widely accepted and their authenticity and adherence to organic agricultural practices are ensured by certifications like Global Organic Textile Standard (GOTS).

Example: The environmentally hazardous effects of traditionally used mordants in natural dyeing processes have been linked to wastewater generated from the dyeing and finishing processes, which contain effluents. As an alternative that is better for the environment than conventional metallic mordants, plant-based mordants have become increasingly popular in recent years. Some plants, known as hyperaccumulators, have the inherent ability to absorb aluminum through their roots and are hence suitable for use as biomordants. Examples of such plants include tea, camelia, and club moss.

Plants that are capable of hyperaccumulation can be utilized as biomordants in the production of natural dyes. These plants can transfer the metal that they have taken up from the surrounding soil into the cloth when they are used in the dyeing process. This eliminates the need for synthetic or metallic mordants and gives an alternative that is both natural and friendly to the environment for binding colors to textiles. The recent advancements in natural dyes in mordants support in various aspects of textile industry (Fig. 5).

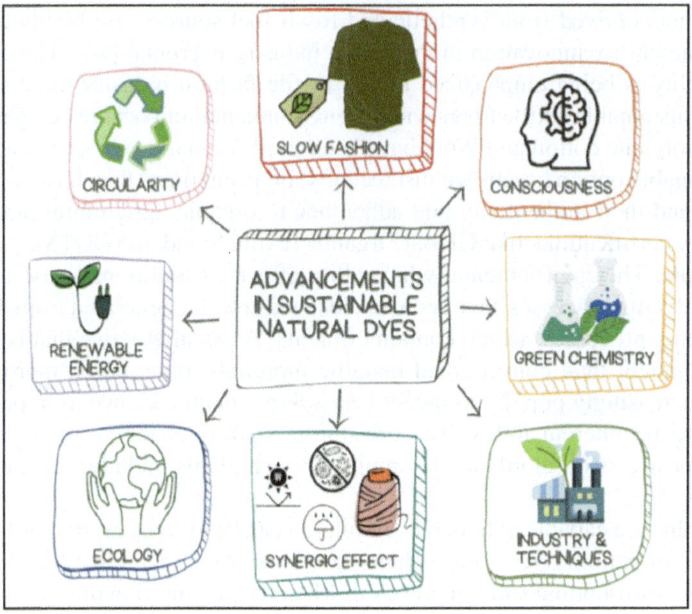

Fig. 5 Advancement in sustainable natural dyes [14]

5 Packaging

The role that clothing packaging plays is crucial in the larger context of the sustainable textile industry. People usually think about the materials used to make clothes and how they are made when they talk about ecology, but packaging should also be taken into account.

Consumption of Resources: The packaging business uses a lot of new paper and plastic, both of which come from natural sources. When it comes to textiles, a significant quantity of these materials is utilized to package various types of apparel. This consumption adds to the depletion of resources as well as the environmental problems connected with the mining and processing of raw materials.

Trash Generation: A sizable portion of all trash is produced by the textile industry as a result of the use of paper and plastic in product packaging. The problem of trash around the globe can be eliminated when discarded goods such as packaging are dumped in landfills [15].

5.1 *Impact of Packaging Materials on the Environment*

According to certain studies, textile packaging accounts for a sizeable fraction of the total amount of paper and plastic that is manufactured worldwide. The environmental impact of this practice is shown by the data that follow, which are as follows:

- Paper used for packaging textiles accounts for over half of the total annual production of paper.
- Approximately forty percent of the plastic that is manufactured each year is used for the packaging of textiles.
- Despite the enormous quantity of packaging materials that are produced, only 36% of all solid waste generated in municipalities can be attributable to the packaging industry across all industries. One big problem is that there isn't enough recycling and waste handling.

The following are characteristics of sustainable packaging according to the Sustainable Packaging Coalition's definition.

Efficient protection and preservation of the product inside the sustainable packaging is essential to its long-term viability. The percentage of used sustainable packaging types is mentioned in Fig. 6. From the extraction and preparation of raw material up to the point of usage by the consumer, it should make sure that the clothes stay in great shape.

Therefore, the features of sustainable packaging are:

1. Beneficial, safe, and healthy throughout its lifecycle.
2. Performance and cost criteria are met.
3. Renewable energy is used for all alternatives.
4. Usage of renewable and recycled materials.
5. Use of clean production technologies.

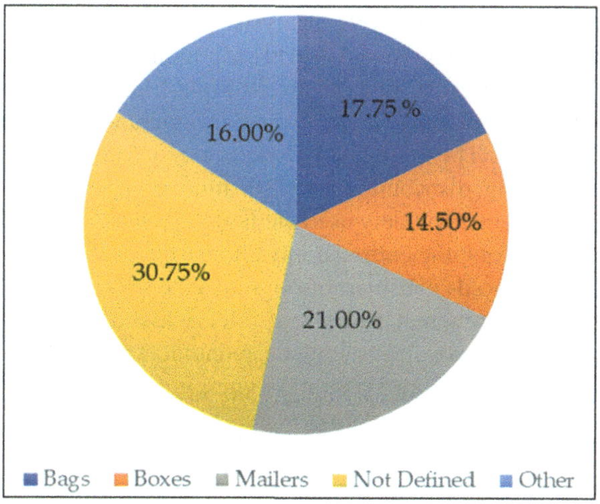

Fig. 6 Percentage of used sustainable packaging types [16]

To fight the environmental catastrophe, more than 120 countries have imposed fees and prohibitions on single-use plastic packaging. Although new government laws are enough to stop the accumulation of packaging trash and pollution.

Therefore, various waste management practices such as

1. Schemes for extended producer responsibility in packaging and durable goods
2. Depending on the nation and industry sectors, there may be landfill taxes, deposit and refund policies, and pay-as-you-throw systems for single-use packaging. It is acknowledged that the selection of a packaging system is significantly influenced by economic, technological, and legal considerations [17, 18].

5.2 Sustainable Packaging

Sustainable packaging is defined as the invention and use of packaging that improves sustainability and has a lower environmental impact. According to industry experts, solutions for environment-friendly packaging should be practical, structured, and profitable. As a result, when considering environmentally sound packaging options, certain sustainability indicators must be taken into account on a priority basis:

For example

- Does the new packaging system help cut down on packaging waste?
- Whether it improves material efficiency, product-to-packaging ratios, and
- Whether it is biodegradable, recyclable, returnable, renewable, or recyclable.

6 Life Cycle Assessment

Other names for life-cycle assessment (LCA) include eco-balance, cradle-to-grave analysis, and life-cycle analysis. It is commonly acknowledged that the textile industry is a significant worldwide industry with a significant impact on the environment at every stage of its multiple life cycle phases. The LCA methodology is widely employed to assess the environmental impact of various stages in the garment life cycle. These stages include raw material procurement and extraction, product processing, distribution and utilization, product disposal, and product recycling. This assessment is a widely used methodology that looks at and assesses the environmental effects along the entire supply chain of the clothing manufacturing process.

LCA holds considerable significance within the textile industry due to the potential environmental and social effects associated with each stage of the industrial process. These stages include manufacturing, which may entail the utilization of toxic chemicals or substantial energy consumption, as well as the final use of the product, whereas certain products emit hazardous substances upon disposal in landfills [19]. Figure 7 depicts the various steps involved in the LCA for the apparel industry.

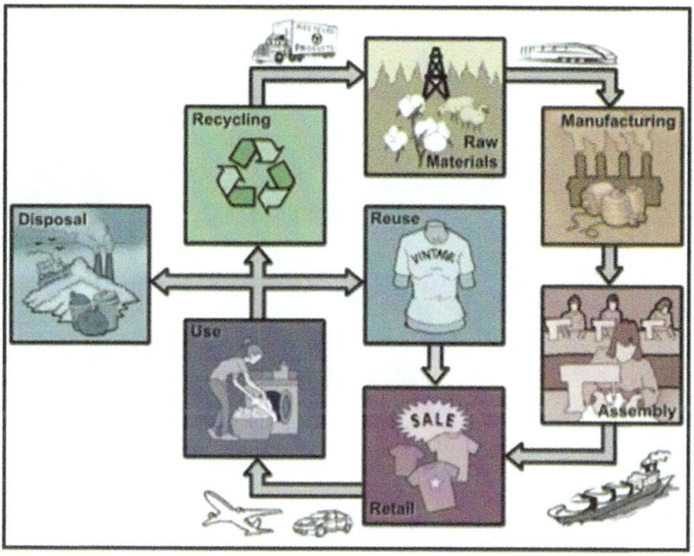

Fig. 7 Steps of life cycle assessment for the apparel industry [20]

7 Process Innovation

Advancement in textile processing sustainability can be achieved by the implementation of various process innovation practices such as:

1. Cleaner production.
2. Alternative dyeing and finishing technologies and machinery (eco-efficiency).
3. Supply chain management.
4. Sustainable technology.
5. Textile processing using enzymes.

7.1 Cleaner Production

The textile industry is responsible for more consumption amount of water. An integrated and deterrent environmental business strategy that increases overall efficiency, reduces risks to people and the environment, and produces higher-quality goods leads to cleaner production (Fig. 8). Therefore, the textile and apparel industry's adoption of improved production techniques helps to reduce environmental impacts and produce highly environmentally friendly outcomes [21]. Numerous novel procedures, including alternative dyeing technologies employing biomass pigment, should be adopted.

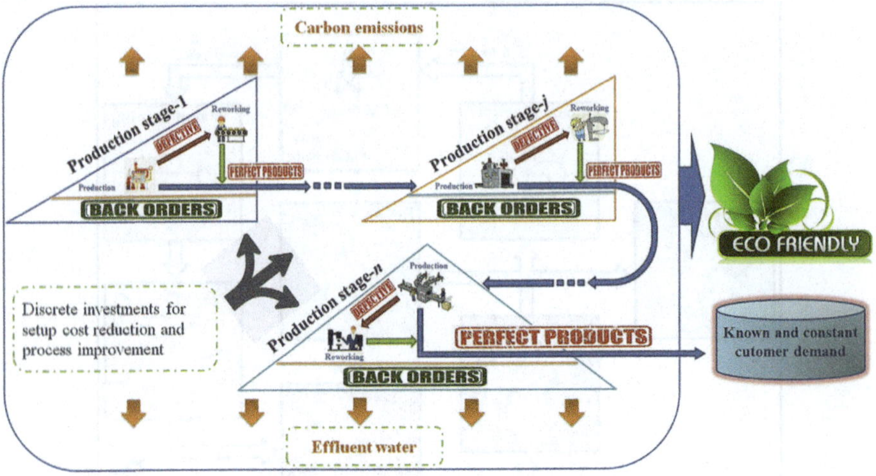

Fig. 8 Optimization of the process for a cleaner production system for sustainable development in textile processes [23]

The United Nations Environment Programme (UNEP), was established in 1972, the goal of the program was: the coordination of global initiatives on environmental protection and the advancement of sustainable development. The objectives of the cleaner production programs, aim to raise international consciousness regarding the implementation of environmental protection strategies and influence industries to adopt these strategies with government support. The 2030 agenda, which was also discussed and clarified at that time, comprises seventeen sustainable development goals with 169 targets. The economic, social, and ecological facets of sustainable development are among these goals. Cleaner production, according to the sustainable development goals, greatly aids in the accomplishment of goals 15 (protect terrestrial life), 12 (sustainable production and consumption), and 9 (industry, innovation, and infrastructure) [22]. Acquisition of cleaner production not only restricts the development of new alternative materials, but also creates opportunities for cluster process management, investment in the purchasing of new equipment, technological advancement, and lifecycle assessment in the textile industry. The more sustainable production approach, also known as a proactive environmental protection strategy, promotes optimization of the various manufacturing processes while taking into mind the environmental impact. Alleviating the adverse environmental impacts resulting from textile effluents, such as the discharge of scouring and bleaching materials, dyes, finishing, and printing chemicals, is very important for the sustainable textile industry [23].

7.2 *Sustainable Dyeing, Finishing Technologies, and Machinery (Eco-efficiency)*

In simpler terms, replacing old machinery and tools with the newest technology has fewer negative effects on the environment. This can only be accomplished through the implementation of various processes that reduce pollution. **For example**, the efficacy was doubled with the implementation of jet-weaving looms in comparison to old technology. Eco-efficiency is defined as the capacity of an organization or firm to produce services and products at competitive prices in the market along with reducing resource consumption and environmental adverse effects throughout their lifetimes. Hence, it is the principle of "generating greater value while minimizing environmental harm." Eco-efficiency is frequently associated with technological advancements.

According to Armstrong research suggests that good pumping systems, automatic dye disposal, chemical disposal, and jet dyeing machines proved to be significant technologies to reduce water pollution in terms of sustainability for the environment. Its low material-to-liquor ratio, use of natural dyes for pollution prevention and control, sophisticated oxidation processes with Fenton's reagent, and membrane bioreactors can all help to maximize resource efficiency. Additionally, supercritical fluid dyeing machines eliminate supercritical carbon dioxide for water in the textile dyeing process, eventually reducing water consumption and eliminating the consumption of chemical fixatives, which ultimately contribute to pollution reduction [24].

- **Irradiation technologies** (Fig. 9), including electron beam irradiation, plasma treatment, and ultrasound and radiation, are increasingly being adopted as sustainable alternatives to traditional wet process finishing methods in the textile dyeing and finishing sector.

7.3 *Supply Chain Management*

SCM is a considerable procedure that is implemented within the textile industry. Its responsibilities extend from obtaining or acquiring of raw materials to the distribution of finished products to the ultimate buyers or consumers. This encompasses the coordination of financial resources, material and information flows, and production and distribution processes. The textile industry's supply chain is of utmost importance owing to its intrinsic intricacy, worldwide scope, and the constantly changing preferences of consumers for an extensive array of textile merchandise. The textile supply chain is characterized by its global nature and intricate structure, which comes from its multifaceted stages encompassing raw material procurement, manufacturing, transportation, and distribution. The complex issues of logistics and coordination are further complicated by the worldwide scope of the textile industry, which frequently entails procuring materials and conducting production in various nations [26].

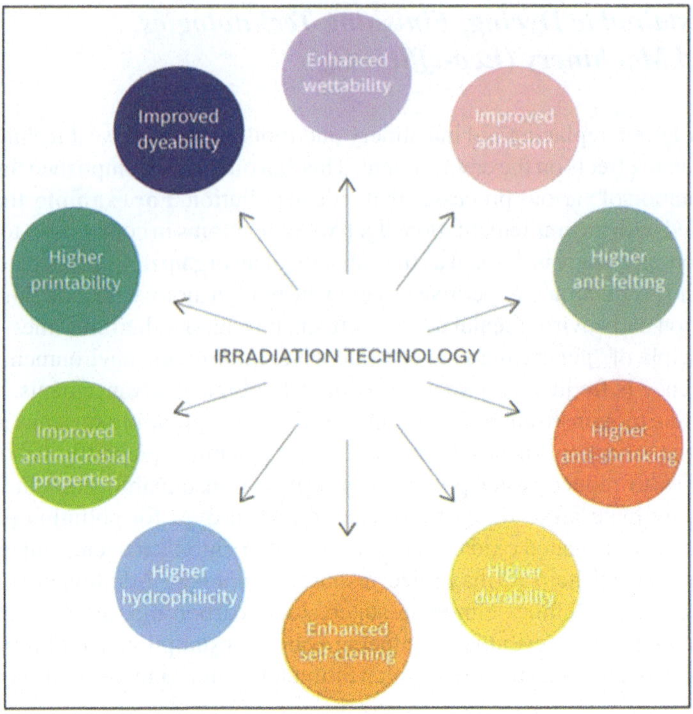

Fig. 9 Effects of irradiation technologies on textile materials [25]

One of the foremost obstacles in managing the textile supply chain is the establishment and maintenance of solid relationships with both downstream (distribution and retail) and upstream (raw material suppliers) processes. These relationships are crucial for ensuring that the flow of materials and products is seamless and effective. It is imperative to cultivate collaborative relationships throughout the supply chain due to the interdependence of all stages. However, this can prove to be a formidable task, especially when managing intricate, worldwide networks [27].

To safeguard the textile industry's sustainability and foster innovation, it is critical to fortify these associations. This entails cultivating operational flexibility that enables prompt adaptations in response to fluctuating demands or disruptions. Additionally, problem-solving and innovative thought are essential for addressing the unique challenges that may arise in the supply chain. Adaptability and creativity are critical attributes in effectively addressing changes in market dynamics, consumer preferences, and global events.

Some key aspects of supply chain management in the textile industry include: Sourcing raw materials, production planning (demand forecasting, capacity planning), ensuring the quality and consistency of textiles through various quality control processes at each stage of production, management of inventory levels to keep in balance the need for availability of the product with the cost of carrying inventory, selecting suitable transportation modes and routes, ensuring that supply chain practices align with sustainability goals such as elimination of carbon footprints and consumption of eco-friendly sources, building strong relationship with supplier through communication and shared goals, ensuring specific compliance with companies specific standards, labeling and regulations, meet customer demands and expectations for quality and responsiveness to inquiries and complaints. For a highly competitive and rapidly changing market, a well-managed supply chain can reduce lead times, lower production costs, and better production quality to meet sustainability in textiles.

7.4 Sustainable Waste Handling Practices

Waste management is a significant concern that requires worldwide attention within the textile industry. Management constitutes a category of sustainable innovation that emphasizes efforts to minimize, repurpose, or recycle materials. Circular economy entails the redesign of the manufacturing process so that resources are utilized in their production, usage, and allocation with an emphasis on recycling and reuse to the greatest extent possible. The recent development in textile industry toward circular economy is depicted in Fig. 10. The textile industry's high consumption of resources such as carbon, water, and oil contributes to its negative societal impact; therefore, the circular economy concept should be taken into consideration for its environmental, industrial, economic, and ecological aspects. The implementation of a circular economy is hindered by numerous obstacles in the fashion industry, such as stylistic preferences and consumer fashion trends.

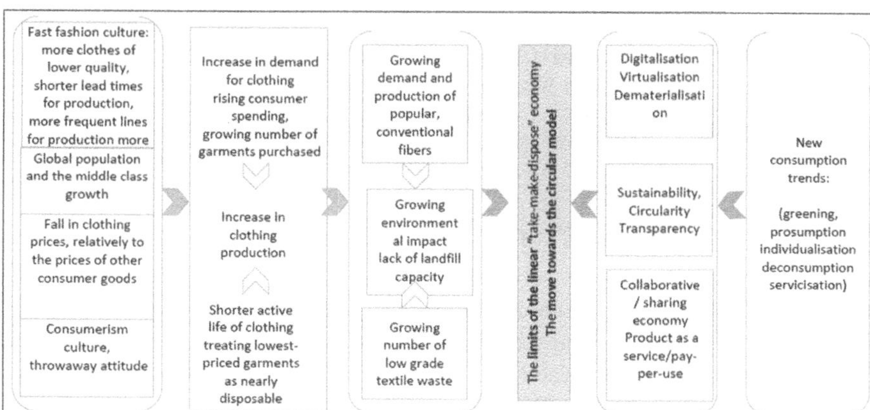

Fig. 10 Latest development in the textile and garment industry transitions toward circular economy [29]

Table 1 Basis for circular economy [28]

S. No.	Principle	Description
1	Design for reuse	Technical and biological products that are developed for reuse in a subsequent cycle will be devoid of refuse throughout their life cycle. Thus It is important to reassemble and utilize the material, ultimately the product's value will be reused
2	Resilience through diversity	To ensure that an economy is resistant to unknown external influences, a realistic approach to business strategy must be diverse
3	Energy use from infinite sources	The additional energy produced during the upcycling procedure may be renewable. The circular economy is predominately concerned with renewable energy rather than labor
4	System thinking	Back-feed loops are of the greatest significance when considering the fundamental principle of a nonlinear system. Significant time investment is necessary across various levels of the production system. At various industry scales, feedback mechanisms are considered a good contributor to the resilience of the circular economy
5	Bio-based basis	Popularly, consumer products are created using biological materials. Naturally scoured products are the most preferable

Strict regulations govern the discharge of effluent water from the textile industry, given the detrimental consequences that endanger both the environment and society at large. Standards for effluent water discharge have been established by the Indian Central Pollution Control Board (CPCB) under unfixed environmental safety requirements and local conditions. The basic principles of the circular economy are described in Table 1.

7.5 Enzymatic Textile Process

Enzymes are regarded as green chemicals because of their biodegradability and production with renewable and recyclable resources. Also these processes include working at lower temperatures and atmospheric pressures, pH, and less consumption of water and energy resources. The advancements and developments in the usage of enzymes in textile industry are mentioned in Fig. 11.

The enzymatic textile process is used in garment and laundry operations throughout the whole multifaceted textile production process, from procurement of raw material to final finished product or service. Enzymes can be utilized in

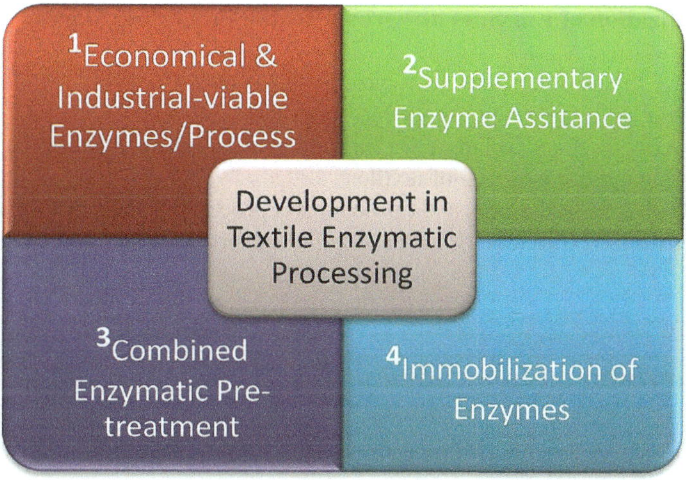

Fig. 11 Advancements in the usage of enzymes for textile processing [30]

every stage of chemical processing, including the removal of effluent from dye and finishing bath, etc. To go with a particular process with enzymes, it is necessary to incorporate them into the process. And finally, it is deactivated and disposed of in effluent. This method of enzyme application is not cost-effective, and enzymes used in textile processing must be solid against harsh parameters such as high temperature, pH, salts, alkalis, and surfactants that are commonly found in the textile industry. Enzymes are thought to be the best substitute for dangerous chemicals and dyes that harm the environment. Enzymes including amylase, cellulase, catalase, protease, pectinase, laccase, and lipase have been used more frequently in the textile manufacturing sector (Fig. 12). The enzymatic method exhibits considerable promise due to its environmentally friendly nature, ability to yield products of good quality, and ability to conserve energy, water, and time.

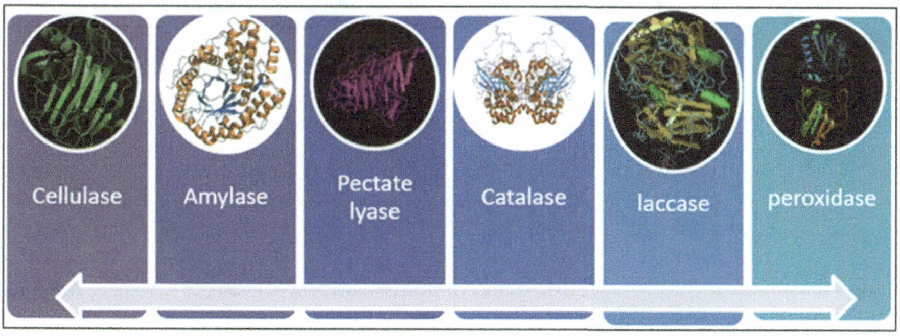

Fig. 12 Different types of enzymes and their microscopic structures [31]

7.5.1 Utilization of Enzymes at Various Textile Processing Stages

A. Amylase, lipase for enzymatic desizing
B. Pectinase, cellulase, cutinase for bio-scouring
C. Oxidoreductase, and xylanase for bleaching and garment washing
D. Oxidoreductase for dyeing of textiles
E. Cellulase, oxidoreductase, and lipase for wet processing and finishing of textiles
F. Composting (textile waste): Laccase, cellulase, protease, nylonase, polyesterase

8 Organizational Innovation

The phrase "organizational innovation" (Fig. 13) encompasses a variety of approaches, including corporate policy, environmental management systems (EMS), collaboration, business practices, culture-knowledge management, and risk management. Organizational innovation contributes to economic development through knowledge sharing and innovation activities within the organization, fostering new markets and improvements in present markets, and connecting with high-level business processes [32]. Successful companies recognize that adopting new management practices that completely or partially fit with their culture increases the success and growth of any business.

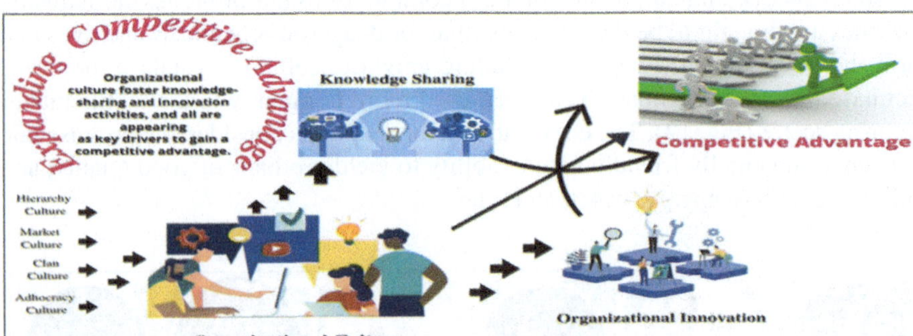

Fig. 13 Organizational innovation, knowledge sharing, and culture for high competitive advantage [32]

8.1 Environmental Operational System (EMS) and Corporate Environmental Strategies

Environmental management in businesses refers to the strategic and operational measures taken by an organization to lessen its adverse effects on the environment. Environmental management practices, which refer to competitive arrangements toward the environment pillar of sustainability and various operational activities with a focus on saving the environment, become evident through an integrated approach, according to "conventional" corporate management.

Environmental management practices (EMP) include all policies and operations governing reducing the organization's effect on the environment due to its business practices. Aside from that, a lot of management business practices are frequently combined into a single "environmental" practice, even though it's vital to very successfully separate practices from one another. Environmental management practices can be internal corporate policies that are not as sophisticated as international certifications like ISO 14001, or they can be structured, organized, optimized, standardized, and widely accepted [33].

8.2 Business Model

The fashion industry is considered the second largest industry at a global level due to the negative effect on the environment because of its operations in supply chain management. Nevertheless, fast fashion, which is defined by lower production costs, frequent consumption, and selective use of specific clothing, is becoming more and more popular despite general awareness of the textile industry's detrimental effects on the environment. Businesses, both new and old, are in search of new ways and techniques to do well in a competitive market by coming up with new ways to do business along with socially responsible and environment-friendly practices. Emergence of some practices such as entrepreneurial ideas that address this issue include low consumerism, fair trade, sharing economy, and circular economy; but there is still a gap between the various levels of environmental and social sustainability [34].

The objective of the new circular economy (Fig. 14) aimed at modifying and bending the continuous economy to more sustainable applications mainly focused on four major types of the clothing system cycle which are as follows: materials, production, usage, and disposal.

Example: Triple bottom line (TBL) (Fig. 15) is an important tool to understand the concept of regenerative and more sustainability aspects applied to companies or businesses with a main focus on economic, environmental, and social responsibilities. Additionally, Triple bottom line has provided a foundation for many companies to develop a sustainable business model.

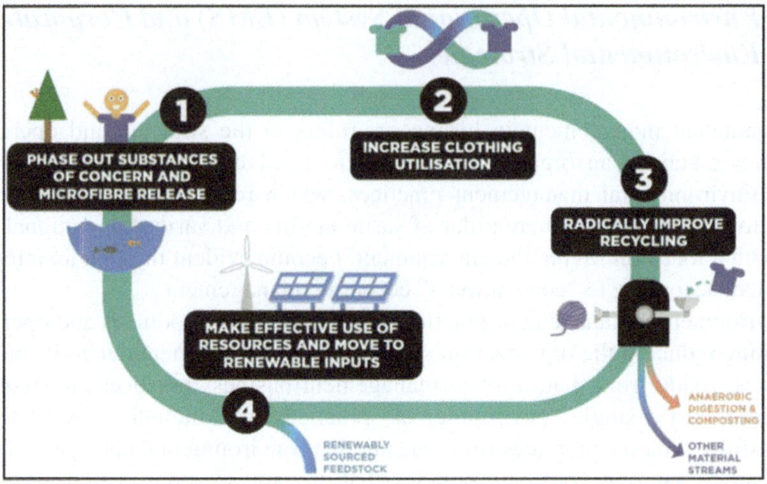

Fig. 14 Objectives of the proposed circular economy for apparel [35]

Fig. 15 Triple bottom line [36]

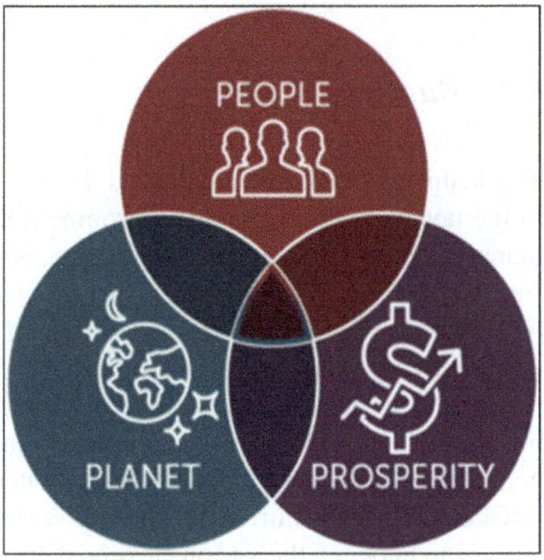

8.3 Collaborations

In India for sustainable development in textiles, prominent business players and organizations have undertaken major ways and processes to introduce sustainability in the textile manufacturing value chain and focused on input management as an important and preferred business strategy. To accelerate the shift to a more sustainable society, cooperation between businesses, consumers, governments, non-governmental organizations, universities, and other businesses is crucial. One of the most economically industrialized industries with sustainability issues is global fashion. Such innovations frequently call for cooperative development, which brings to light several issues related to stakeholder collaboration. By fostering acceptance of sustainable technologies and practices for waste reduction and enhancing the environmental and social performance of any business or organization, collaboration enhances sustainable benefits. In sustainable supply chain management practices, the institutional environment—that is, environmental laws and regulations in general—has a positive impact on cooperation and good relationship management. Significant advantages of collaboration include the sharing of resources, facilities, and management skills, which eventually allow for more systemic changes in the industry [7].

Collaboration within two or more independent firms is considered one of the core and important governance structures that reciprocate relations between firms, markets, and hierarchies. For pursuing a sustainable strategy, collaboration within two or more independent firms was proved to be an important governance structure.

8.4 Culture and Knowledge Management

KM (knowledge management) refers to the process of identifying, creating, storing, disseminating, and application of knowledge in organizations and firms. In the past several years it has gained more attention in practice and research. For organizational growth and success, there are cultural traits that are responsible for strengthening knowledge sharing and capabilities.

To continue competitive advantage and to upgrade customer satisfaction, Knowledge should be learned, shared, and applied [37]. To endure and change in functioning sustainability as well as for competitive advantage for shareholders and also for customer's trust, knowledge management practices should be taken into consideration [38], which has been acknowledged as an important component in creating and disseminating knowledge for developing new services and innovative products.

According to some authors, corporate sustainability is a key driven by combined external and internal forces, and flexibility is the most suitable approach to address sustainability challenges. It was discovered that there is a significant correlation between green/sustainable technological usage and environmental awareness when it comes to knowledge management practices. This correlation is influenced by the behavior of the organization's business leaders as well as the organization's culture, which is advantageous for the organization's long-term ability to maintain high-performance levels.

9 Conclusion

Today's textile and garment industry mainly relies on fast cycles of fashion trends with the main focus to continuously produce products according to consumer needs and demands simultaneously saving the natural environment and energy without affecting the requirements and satisfaction of consumers. Textile production is considered one of the most polluting production processes as it has experienced growth and success in the last few years. Learning about sustainable innovations in the textile industry was the goal of this review. Product innovation related to sustainability includes ecodesign, eco-label, life cycle assessment, sustainable materials, and sustainable packaging; process innovation includes eco-efficiency, cleaner production processes, eco-wise waste management, supply chain management, and enzyme-based textile processing; and organizational innovation includes environmental management systems (EMS), the development, implementation, monitoring, and upkeep of organizational policies and procedures, cooperation between businesses and consumers, business model innovation, culture, and knowledge management. The fashion industry has developed a unique set of sustainability-focused technologies, such as enzymatic textile processing for the sector as a whole and eco-labeling for individual products. The implementation of these concepts, however, may be complicated by several variables, including cost, knowledge shortages, support from the government, and broader economic issues. Conquering these obstacles is necessary to promote environmentally responsible practices within the textile industry, which is known for the harm that it does to the natural environment.

References

1. Palomares I, Martinez-Camara E, Montes R, Garcia-Moral P, Chiachio M, Chiachio J, Herrera F et al (2021) A panoramic view and Swot analysis of artificial intelligence for achieving the sustainable development goals by 2030: Progress and prospects. Appl Intell 51:6497–6527
2. Change NC (2018) The price of fast fashion. Nat Clim Change 8:1–1
3. Niinimaki K, Peters G, Dahlbo H, Perry P, Rissanen T, Gwilt A (2020) The environmental price of fast fashion. Nat Rev Earth Environ 1(4):189–200
4. Shirvanimoghaddam K, Motamed B, Ramakrishna S, Naebe M (2020) Death by waste: fashion and textile circular economy case. Sci Total Environ 718:137317
5. Abbas S, Halog A (2021) Analysis of Pakistani textile industry: recommendations towards circular and sustainable production. Circular Econ 77–111. https://doi.org/10.1007/978-981-16-3698-1_3
6. Salo HH, Suikkanen J, Nissinen A (2020) Eco-innovation motivations and ecodesign tool implementation in companies in the Nordic textile and information technology sectors. Bus Strategy Environ 29:2654–2667
7. Gu C, Xu C, Zhou Q, Shen C, Ma C, Liu S, Yin S, Li F (2021) Congener and isomer specific per fluorinated compounds in textile wastewater from Southeast China. J Clean Prod 320:12889
8. Rubik F, Frankl P (2017) The future of eco-labelling: making environmental product information systems effective. Routledge
9. Russel AH, Robidas LC (2019) Opportunities and challenges of eco-label practices in Bangladesh for promoting environmentally conscious consumers. Eur J Bus Manage Res 4(6). https://doi.org/10.24018/ejbmr.2019.4.6.158
10. Eryuruk SH (2012) Greening of the textile and clothing industry. Fibres Text Eastern Europe 95:22–27
11. Ranasinghe L, Jayasooriya VM (2021) Ecolabelling in textile industry: a review. Resour Environ Sustain 6:100037
12. https://www.ecolabelindex.com/ecolabels/

13. Adu C, Zhu C, Jolly M, Richardson RM, Eichhorn SJ (2021) Continuous and sustainable cellulose filaments from ionic liquid dissolved paper sludge nanofibres. J Clean Prod 280:124503
14. Pizzicato B, Pacifico S, Cayuela D, Mijas G, Riba-Moliner M (2023) Advancements in sustainable natural dyes for textile applications: a review. Molecules 28(16):5954
15. Holding A (2019) Polybags in the fashion industry: evaluating the options. Good Fashion 3–13
16. Jestratijevic I, Vrabic-Brodnjak U (2022) Sustainable and innovative packaging solutions in the fashion industry: global report. Sustainability 14(20):13476
17. Herbes C, Beuthner C, Ramme I (2020) How green is your packaging—a comparative international study of cues consumers use to recognize environmentally friendly packaging. Int J Consum Stud 44:258–271
18. Silva N, Palsson H (2022) Industrial packaging and its impact on sustainability and circular economy: a systematic literature review. J Clean Prod 333:130165
19. Klewitz J, Hansen EG (2014) Sustainability oriented innovation of SMEs: a systematic review. J Clean Prod 65:57–75
20. https://oroeco.wordpress.com/2012/09/15/lca-life-cycle-assessment/
21. Ibrahim NA, Khalil HM, Eid BM (2015) A cleaner production of ultra-violet shielding wool prints. J Clean Prod 92:187–195
22. Schellenberger S, Hill PJ, Levenstam O, Gillgard P, Cousins IT, Taylor M, Blackburn RS (2019) Highly fluorinated chemicals in functional textiles can be replaced by re-evaluating liquid repellency and end-user requirements. J Clean Prod 217:134–143
23. Tayyab M, Jemai J, Lim H, Sarkar B (2020) A sustainable development framework for a cleaner multi-item multi-stage textile production system with a process improvement initiative. J Clean Prod 246:119055
24. Armstrong CM, Niinimäki K, Kujala S, Karell E, Lang C (2015) Sustainable product-service systems for clothing: exploring consumer perceptions of consumption alternatives in Finland. J Clean Prod 97:30–39
25. Zhang RP, Cai ZS (2011) Study on the natural dyeing of wool modified with enzyme. Fibers Polym 12:478–483
26. UNOPS. A guide to environmental labels-for procurement practitioners of the United Nations system. UNOPS, Copenhagen
27. Olatubosun P, Charles E, Omoyele T (2021) Rethinking luxury brands and sustainable fashion business models in a risk society. J Des Bus Soc 7:49–81
28. Rathinamoorthy R (2019) Circular fashion. Circular Econ Text Apparel 13–48. https://doi.org/10.1016/b978-0-08-102630-4.00002-9
29. Koszewska M (2018) Circular economy—challenges for the textile and clothing industry. Autex Res J 18(4):337–347
30. Madhu A, Chakraborty JN (2017) Developments in application of enzymes for textile processing. J Clean Prod 145:114–133
31. Kiron MI (2021) Application of enzymes in textile chemical industry. Textile learner. https://textilelearner.net/application-of-enzymes-in-textile/
32. Azeem M, Ahmed M, Haider S, Sajjad M (2021) Expanding competitive advantage through organizational culture, knowledge sharing and organizational innovation. Technol Soc 66:101635
33. Resta B, Dotti S, Boffelli A, Gaiardelli P (2015) Environmental management practices for the textile sector. In: IFIP international conference on advances in production management systems, vol 459. Springer, Cham, pp 625–631
34. Todeschini BV, Cortimiglia MN, Callegaro-de-Menezes D, Ghezzi A (2017) Innovative and sustainable business models in the fashion industry: entrepreneurial drivers, opportunities, and challenges. Bus Horiz 60(6):759–770
35. Morlet A, Opsomer R, Herrmann S, Balmond L, Gillet C, Fuchs L (2017) A new textiles economy: redesigning fashion's future. Ellen MacArthur Found 1–150
36. UW Extended Campus (2022) An explanation of the triple bottom line. https://uwex.wisconsin.edu/stories-news/triple-bottom-line/

37. Li F, Lu H, Hou M, Cui K, Darbandi M (2021) Customer satisfaction with bank services: the role of cloud services, security, e-learning and service quality. Technol Soc 64:101487
38. Nayak B, Bhattacharyya SS, Krishnamoorthy B (2021) Strategic advantage through social inclusivity: an empirical study on resource based view in health insurance firms in India. J Clean Prod 298:126805

The Conclusion: Overview of the Importance of Eco Textiles and the Commitment to Be Made to Have a Sustainable Environment

Sadhna, S. Greeshma, and Rajesh Kumar

The textile industry can have a significant detrimental effect on society and the environment in addition to its enormous economic impact on the world economy. The textile industry's growing recognition of the value of environmental conservation is demonstrated by the emergence of sustainable textiles. By employing eco-friendly materials, cutting waste, and making better use of natural resources, sustainable textile production seeks to lessen its negative effects on the environment. We can lessen pollution, slow down the natural resources' depletion, and lessen the negative effects of fast fashion by selecting sustainable textiles [1].

Approximately 20% of all clean water pollution worldwide is caused by the use of dyes and finishing products used in the textile industry, which makes textile production a major contributor to environmental problems. The effects of washing synthetic fabrics go beyond contaminated water; the ocean floor now contains over 14 million tonnes of microplastics from textile washing. The widespread pollution that comes from the clothing industry has a negative impact on the ecosystems, wildlife, and local communities that are close to these manufacturing facilities.

An estimated 10% of the world's carbon emissions are attributed to the fashion industry, which is a major contributor to this environmental crisis and emits more carbon than both international travel and maritime shipping put together. Furthermore, the textile industry itself emits a significant amount of CO_2, surpassing the environmental effects of sectors like international shipping and aviation. Among the aspects of the industry that harm the environment, the most is the production of textiles, both natural and synthetic.

The textile waste crisis, which is characterized by significant quantities ending up in landfills owing to low recycling rates, exacerbates the environmental cost. This problem is made worse by the textile dyeing process, which releases untreated

Sadhna · S. Greeshma (✉) · R. Kumar
Department of Fashion Design, School of Arts and Design, Woxsen University, Hyderabad, Telangana 502345, India
e-mail: greeshma.s_phd.2023@woxsen.edu.in

Sadhna et al. (eds.), *Climate Action Through Eco-Friendly Textiles*, SDGs and Textiles,
https://doi.org/10.1007/978-981-99-9856-2_13

wastewater into nearby water systems, introducing hazardous materials and heavy metals that endanger the health of animals and people living nearby. Moreover, wildlands must be destroyed to extract fossil fuels for textile manufacturing, which fragments habitat and interferes with important animal behaviours. The pressing necessity of sustainable alternatives and ethical practices to lessen the textile industry's significant environmental impact is highlighted by the state of the industry today [2].

Eco-friendly textiles are a significant development in the textile industry since they prioritise ecologically friendly methods over the use of chemicals and pesticides during the cultivation process. The importance of creating environmentally friendly textiles stems from the textile industry's widespread involvement in landfill waste, greenhouse gas emissions, and water pollution. It is a significant source of CO_2 emissions, outpacing several industries, such as shipping and international aviation. Eco-friendly textiles make use of naturally occurring materials that can be recycled or biodegraded. This calculated decision actively attempts to reduce the pollution footprint of the textile manufacturing industry in addition to being in line with sustainability objectives.

The natural ability of eco-friendly textiles to withstand mould and their hypoallergenic qualities make them a great option for people who are allergic to most things. In addition to their health advantages, these textiles offer consumers stylish options without sacrificing aesthetics by skillfully fusing environmental responsibility with style. The strength and resilience of eco-friendly textiles add even more allure. Customers who adopt these textiles support sustainable living practices in addition to making a fashion statement. The increasing demand for these textiles highlights the need for products that balance ecological well-being and reflect a changing consumer consciousness. Eco textiles are essentially the embodiment of a responsible approach to the textile industry, showing that strength, style, and sustainability can all coexist peacefully in the fabric of our choices [3].

Eco-friendly textiles have a big effect on the world economy. The expansion of green industries is facilitated by the adoption of sustainable practices in the textile industry. It generates employment in the recycling, renewable energy, and sustainable agriculture sectors. Sustainable textiles also lessen their impact on the environment, saving money on water and waste management. The economy may then benefit from these savings. One study estimates that by 2030, resolving the environmental and social problems caused by the fashion industry would add \$192 billion to global GDP. Every year, garments worth more than \$400 billion are thrown away before their time [4].

Natural fibres are a renewable and biodegradable resource that comes from either plants or animals. They are an environmentally friendly substitute that is sustainable. As a result of their remarkable strength, stiffness, durability, and lack of toxicity, natural fibres are becoming more and more popular than their synthetic counterparts. Natural fibres are less dense than mineral fibres, and because they require the extraction of raw materials, their production has a comparatively smaller environmental impact. This environmentally friendly trait is especially notable in the modern setting, where the move towards sustainable practices is motivated by environmental concerns.

Traditional natural fibres include viscose, polyester, nylon, cotton, acrylic, long vegetable fibres, and polyester. Important fibres such as cotton, flax, and jute are found in the vegetable or cellulose-based category, whereas wool, mohair, and silk are found in the animal or protein-based category. Among the mineral class, asbestos stands out as a significant fibre.

Alongside these traditional choices, a fresh wave of natural fibres is starting to surface. Crafted from biodegradable or recyclable materials, these fibres contribute to reducing pollution in the textile industry. As an illustration of such novel fibres, consider coir, bamboo, and akund floss. Notable natural fibres with better mechanical qualities include hemp, jute, sisal, bamboo, and coir, which make them competitive alternatives to synthetic materials. Natural fibres have changed over time, reflecting a greater dedication to sustainable materials and efforts to reduce environmental impact and improve the textile industry's eco-friendliness [5].

It is remarkable how natural fibres can help mitigate the effects on the environment. Compared to synthetic fibres, natural fibres have a lower carbon footprint throughout their life cycle. Significant volumes of greenhouse gases are released during the production of synthetic fibres, which exacerbates climate change. On the other hand, natural fibres act as carbon sinks; they take in carbon dioxide from the atmosphere as they grow and hold onto it even after they are made into clothing.

Natural fibres typically have a smaller overall environmental impact than synthetic fibres. In contrast to the chemical reactions and petroleum extraction involved in the production of synthetic fibres, the cultivation and harvesting of natural fibres usually entail less energy-intensive processes. In addition to producing greenhouse gases, the production of synthetic fibres pollutes the air and water. Understanding the advantages of natural fibres for the environment highlights how they support sustainability and lessen the ecological impact of textile production, which is in line with the international need to combat climate change and environmental degradation [6].

Numerous chemicals are used extensively in the textile industry for pre-preparatory processes such as scouring, bleaching, mercerizing, and de-sizing, among others. Sadly, there are serious risks associated with these chemicals for both human health and the environment. Even though they are necessary for the textile industry, materials like dyes and auxiliary chemicals used in the wet processing of textiles raise health and environmental issues. Notably, emissions into water sources are the primary cause of environmental challenges [7].

These chemicals are harmful when they come into contact with the skin or are inhaled, which puts both the environment and workers at risk. Preventative actions should be given top priority to reduce exposure and toxicity to address these issues. The discharge of hazardous gases into the environment, especially during processes like bleaching and dyeing, is a serious problem that could have detrimental effects on both human and animal health. The urgent need for more sustainable practices and alternatives within the textile industry is highlighted by the recognition of the risks to the environment and public health posed by these processes. These efforts are in line with larger initiatives to lessen the negative environmental effects and enhance the safety of textile manufacturing [8, 9].

Accepting environmentally friendly substitutes like enzymes and bioproducts offers a workable plan for reducing the harmful effects of chemicals used in textile processing. In particular, enzymes prove to be efficient replacements, particularly during the preparatory stages. Notwithstanding obstacles pertaining to financial and technical implementation, these replacements are expected to become widely successful due to continuing technological advancements and a growing commitment to environmental stewardship.

Enzymes reduce energy and water consumption and the overall number of chemicals used in textile processing, among other benefits. Enzymes have benefits for the environment in addition to improving the quality of the finished product. The textile industry can help reduce its environmental impact overall and align with sustainability goals by implementing these eco-friendly auxiliaries. The use of enzymes in the textile manufacturing process presents a viable way to address the ecological and technical issues that arise as progress continues and the need for environmental responsibility grows [7].

Five million people's worth of water is used annually by the textile industry, which is one of the biggest users of water resources. It uses an enormous 93 billion cubic metres of water annually. The production of synthetic fabric, which uses 70 million barrels of oil a year, exacerbates environmental problems by contaminating water sources with lead, arsenic, and benzene. Approximately 20% of the wastewater produced worldwide is a result of fabric dyeing and treatment processes, which greatly contribute to water pollution worldwide. An astounding 1500 billion litres of water are used annually by garment factories and mills in Bangladesh alone, which depletes the region's groundwater supplies and introduces dangerous pollutants into adjacent water sources [10].

Water is contaminated by a variety of chemicals during the textile finishing and dyeing processes, including phenol, oil, dyes, pesticides, and heavy metals like chromium, copper, and mercury. After entering adjacent streams and groundwater, this contaminated water threatens food sources by irrigating crops with chemicals that cause cancer. Though less obvious, microfibre pollution from the textile industry is a dangerous threat to water sources. These tiny synthetic fibres can disturb underwater ecosystems because they take hundreds of years to break down. Fish and other marine animals have been found to contain synthetic microfibers in their stomachs, highlighting the extensive environmental effects of water pollution from the textile industry. To address these issues, a dedicated effort must be made to implement sustainable practices, cut back on water use, and create creative solutions to stop the industry's damaging effects on the world's water supplies [10, 11].

The dyeing process involved in the textile industry uses synthetic dyes, which present serious environmental risks. Synthetic dyes are extremely toxic and may cause cancer, endangering the health of both humans and animals. As a result, disposing of them as hazardous waste adds to environmental problems. In addition, the manufacturing process of synthetic fabrics uses an impressive 70 million barrels of oil per year and releases dangerous pollutants into water sources, including benzene, arsenic, and lead.

The harmful effects also extend to wastewater, where practices related to fabric dyeing and treatment produce an astounding 20% of the world's wastewater. This polluted water may make its way into neighbouring groundwater and streams, where it may end up in agricultural irrigation systems. This contamination therefore presents a risk of introducing carcinogenic chemicals into the food chain, thereby establishing a dangerous connection between the safety of our food sources and the practices of the textile industry. The environmental and health risks related to industrial dyeing in the textile industry require immediate attention and viable solutions [12].

Natural dyeing is extremely important in the current global context. Originating from plant materials such as fruits, roots, leaves, and flowers, these dyes have a millennium-long history, but their use dwindled when synthetic dyes were developed in the late 1800s. The growing demand for sustainable and ethical fashion is in line with the renewed interest in natural dyeing techniques, which are driven by a shared commitment to reducing the environmental impact of textile production. Growingly conscious of the environmental impact of their purchases, consumers look for products composed of natural and eco-friendly materials.

Natural dyes have many benefits over their synthetic counterparts, making them a compelling substitute. Because of their biodegradability, they guarantee a low environmental impact, which promotes a sustainable method of producing textiles. Natural dyes are non-toxic and do not endanger aquatic life or human health, in contrast to synthetic dyes. Their sustainability is further highlighted by their sourcing from renewable resources, which supports an ethical approach to the fashion industry.

In addition, natural dyes are attractive because they can create distinct and beautiful hues that are superior to those made with synthetic materials, giving designers a wide range of visually appealing colour options. The resurgence of natural dyeing methods is a significant and progressive step towards greater sustainability in the fashion industry, in line with environmentally conscious consumerism and responsible consumption [13, 14].

To prevent an excessive loss of natural resources, textile production must optimise resource consumption. Eco-friendly textiles use fewer resources and energy. The use of environmentally harmful materials and sustainable printing techniques reduces the need for energy and water use that would otherwise be prohibitive. The textile sector is susceptible to high rates of energy loss in a variety of production processes, which raises energy costs and reduces productivity. Energy-intensive singeing procedures are used to remove singes from the fabric's surface. The textile industry pays between 40 and 45% of its total utility bills, on average, from its electricity bills [3].

The severe lack of energy and water presents a significant challenge to the global textile industry. It is crucial to use water and energy resources as efficiently as possible and to switch to ecologically friendly production techniques. The textile industry can lessen its carbon footprint and help the development of green industries by conserving energy and water. As a result, jobs are created in the recycling, renewable energy, and sustainable agriculture sectors.

The world's textile industry is thought to be the most detrimental to the environment. The environmental issues facing the textile industry are related to the disposal of textiles after use and arise during certain production processes. Formaldehyde, pesticides, carcinogenic dyes, skin neutrality, heavy metal content, pH, and fastness to perspiration are related factors for eco-standards.

Minimalism and capsule wardrobes are crucial for minimizing the use of resources in textile production. A minimalist approach to fashion, capsule collections consist of a limited number of adaptable items that can be combined to create a multitude of looks. This methodology mitigates the necessity for excessive resource consumption while advancing sustainable fashion. Conversely, minimalism is a way of life that emphasises consuming fewer resources and living with less. People can lessen their carbon footprint and help preserve natural resources by leading a minimalist lifestyle.

One of the most significant aspects of the textile industry is the segmentation of the global eco-textile market. By the end of 2030, the global market for eco-friendly textiles is expected to have grown from an estimated USD 6.5 billion in 2022 to USD 13.93 billion [15].

One of the biggest developments in the textile industry is the growing focus on eco-friendliness and sustainability. Demand for textiles made from organic and recycled materials is rising as consumers become more environmentally conscious. This change forces textile producers to use environmentally friendly production techniques and lower their carbon footprint.

The expansion of green industries is facilitated by the adoption of sustainable practices in the textile industry. It generates employment in the recycling, renewable energy, and sustainable agriculture sectors. Sustainable textiles also lessen their impact on the environment, saving money on water and waste management. The economy may then benefit from these savings. It is estimated that the global textile market is worth $1.4 trillion and supports over 300 million jobs. The global textile and apparel industry is estimated to be valued at over $3 trillion and plays a significant role in supporting economies worldwide [16, 17].

When it comes to managing textile waste, the 3Rs—Reduce, Reuse, and Recycle—are essential to safeguarding the environment. Globally, the 3Rs concept is pushed as a means of effectively using resources and materials to create a sustainable material-cycle society. The 3Rs idea can aid in lowering the quantity of waste produced by the textile industry, which produces a lot of waste. The 3R strategy's elements—Reduce, Reuse, and Recycle—can be used to manage textile waste. Ecological production's primary objective is to adopt and put into practice specific strategies to minimise waste generation. To cut down on waste production, the 3Rs concept should be implemented in the clothing and textile industries [18, 19].

Reducing the quantity of fabric used in production, recycling textile waste, and reusing fabric scraps are all ways that the textile industry can cut waste. By implementing sustainable practices like employing eco-friendly materials, conserving energy and water, and using fewer dangerous chemicals, the textile industry can also cut waste. Recycling textiles improves both the environment and the economy by lowering the need for textile chemicals, the amount of land needed for landfills, and the amount of energy and water wasted.

The government has launched a comprehensive programme to encourage environmentally friendly practices in the textile manufacturing industry. This programme takes a multimodal approach, including consumer guidelines and manufacturer regulatory frameworks, with the ultimate goal of fostering an eco-friendly culture.

The government has established a set of guidelines and certifications for textile producers in an effort to encourage and verify environmentally friendly operations. They consist of applying eco-labelling, following the Global Organic Textile Standard (GOTS), meeting Bluesign requirements, and adhering to Oeko-Tex Standard 100. By acting as benchmarks, these certifications help manufacturers understand how important it is to incorporate environmentally friendly and sustainable practices into their production processes.

In addition, consumers still have a part to play in encouraging an environmentally friendly environment. In an effort to promote behaviors that support a more sustainable ecosystem, the government has put out standards and regulations aimed at consumers. These recommendations cover adopting generally sustainable consumption practices, cutting carbon footprints, and managing waste responsibly. The government hopes to empower consumers to make decisions that support environmental conservation objectives by outlining these expectations.

The proposed certifications and regulations are essential for guiding the textile sector toward environmental responsibility. The government hopes to build a framework that not only encourages eco-friendly practices in the textile industry but also protects the environment from the possible harm that unrestrained industrial activity could cause by setting clear expectations and benchmarks. This all-encompassing strategy essentially aims to create a mutually beneficial relationship between the textile sector and the environment, encouraging sustainability and conscientious consumption [20].

The growing concern for environmental sustainability has led to an increasing role for eco-friendly textiles in the textile industry. Eco-friendly textiles are produced using sustainable resources, encourage ethical labor practices, and are made from renewable materials. A commitment to environmentally friendly practices is necessary for the creation of a sustainable environment. Examples of these practices include the adoption of sustainable practices by the textile industry, the application of the 3Rs concept in waste management, the endorsement of eco-friendly brands and businesses, the selection of sustainable textiles, and the dissemination of information about the significance of environmental sustainability [21, 22].

As concerns about environmental sustainability grow, the textile industry is witnessing a significant shift in favor of eco-friendly textiles. These textiles are carefully crafted from sustainable materials, made with eco-friendly methods, and support ethical labor standards. Developing eco-friendly behaviors is essential to maintaining a sustainable environment. This pledge includes implementing sustainable practices in the textile sector, incorporating the reduce, reuse, and recycle (3Rs) waste management principles, endorsing environmentally conscious brands, choosing sustainable fabrics, and actively raising awareness of the critical significance of environmental sustainability.

Individuals and industries work together to create a world that is more environmentally conscious and sustainable by advocating for and putting these practices into practice. This change is a critical step in creating a harmonious balance between industrial operations and protecting the environment for coming generations.

References

1. Chomsky R (2023) The must-have sustainable fabrics for eco-friendly fashion trends. Sustainable review. https://sustainablereview.com/the-must-have-sustainable-fabrics-for-eco-friendly-fashion-trends/
2. The impact of textile production and waste on the environment (infographics): News: European parliament. The impact of textile production and waste on the environment (infographics) | News | European Parliament (2023, June 5). https://www.europarl.europa.eu/news/en/headlines/society/20201208STO93327/the-impact-of-textile-production-and-waste-on-the-environment-infographics
3. Emerging trends revolutionizing the textile sector & their global economic impact. Textile Magazine, Textile News, Apparel News, Fashion News (2023, Oct 5). https://textilevaluechain.in/news-insights/emerg-trends-revolutionizing-the-textile-sector-their-global-economic-impact/
4. Reichart E, Drew D (2019) By the numbers: the economic, social and environmental impacts of "fast fashion." World Resources Institute. https://www.wri.org/insights/numbers-economic-social-and-environmental-impacts-fast-fashion
5. Thyavihalli Girijappa YG, Mavinkere Rangappa S, Parameswaranpillai J, Siengchin S (2019) Natural fibers as sustainable and renewable resource for development of eco-friendly composites: a comprehensive review. Front Mater 6. https://doi.org/10.3389/fmats.2019.00226
6. Gonzalez V, Lou X, Chi T (2023) Evaluating environmental impact of natural and synthetic fibers: a life cycle assessment approach. Sustainability 15(9):7670. https://doi.org/10.3390/su15097670
7. Azanaw A, Birlie B, Teshome B, Jemberie M (2022) Textile effluent treatment methods and eco-friendly resolution of textile wastewater. Case Stud Chem Environ Eng 6:100230. https://doi.org/10.1016/j.cscee.2022.100230
8. Wilson J (2022) Toxic textiles: the chemicals in our clothing. Earth day. https://www.earthday.org/toxic-textiles-the-chemicals-in-our-clothing/
9. Parvin F, Islam S, Urmy Z, Ahmed S, Islam AS (2020) A study on the solutions of environment pollutions and worker's health problems caused by textile manufacturing operations. Biomed J Sci Tech Res 28(4). https://doi.org/10.26717/bjstr.2020.28.004692
10. Bandera G (2023) How the fashion industry pollutes our water. FairPlanet. https://www.fairplanet.org/story/how-the-fashion-industry-pollutes-our-water/
11. O'Donnell D (2022) Water waste in the textile industry & environmental impacts. Sensorex liquid analysis technology. https://sensorex.com/water-waste-textile-industry-environmental-impacts/
12. Lellis B, Fávaro-Polonio CZ, Pamphile JA, Polonio JC (2019) Effects of textile dyes on health and the environment and bioremediation potential of living organisms. Biotechnol Res Innov 3(2):275–290. https://doi.org/10.1016/j.biori.2019.09.001
13. Elsahida K, Fauzi AM, Sailah I, Siregar IZ (2019) Sustainability of the use of natural dyes in the textile industry. IOP Conf Ser Earth Environ Sci 399(1):012065. https://doi.org/10.1088/1755-1315/399/1/012065
14. Hasan MM (2020) Natural dyes: dyeing process and environmental impact. Textile blog. https://www.textileblog.com/natural-dyes-dyeing-process-and-environmental-impact/
15. Virtue Market Research (2023) Global eco-friendly textiles market size, overview, trends, and forecast 2023–2030. https://virtuemarketresearch.com/report/eco-friendly-textiles-market
16. Schumacher KA, Forster AL (2022) Textiles in a circular economy: an assessment of the current landscape, challenges, and opportunities in the United States. Front Sustain 3. https://doi.org/10.3389/frsus.2022.1038323

17. Pacini H (2021) Seizing the opportunities of a circular economy in textiles. UNCTAD. https://unctad.org/news/seizing-opportunities-circular-economy-textiles
18. Kiron MI (2022) Waste management in textile and garment industry. Textile learner. https://textilelearner.net/waste-management-in-textile-and-fashion-industry/
19. Parven S (2023) Waste management in fashion and textile industry: 3R concept, zero waste concept, current trends & circular economy. Textile focus. https://textilefocus.com/waste-management-in-fashion-and-textile-industry-3r-concept-zero-waste-concept-current-trends-circular-economy/
20. Fibre2Fashion (2005) Eco-friendly textiles. https://www.fibre2fashion.com/industry-article/76/eco-friendly-textiles
21. Uekert T, Singh A, DesVeaux JS, Ghosh T, Bhatt A, Yadav G, Afzal S, Walzberg J, Knauer KM, Nicholson SR, Beckham GT, Carpenter AC (2023) Technical, economic, and environmental comparison of closed-loop recycling technologies for common plastics. ACS Sustain Chem Eng 11(3):965–978. https://doi.org/10.1021/acssuschemeng.2c05497
22. Tanveer M, Khan SA, Umar M, Yu Z, Sajid MJ, Haq IU (2022) Waste management and green technology: future trends in circular economy leading towards environmental sustainability. Environ Sci Pollut Res 29(53):80161–80178. https://doi.org/10.1007/s11356-022-23238-8

GPSR Compliance

The European Union's (EU) General Product Safety Regulation (GPSR) is a set of rules that requires consumer products to be safe and our obligations to ensure this.

If you have any concerns about our products, you can contact us on ProductSafety@springernature.com

In case Publisher is established outside the EU, the EU authorized representative is:

Springer Nature Customer Service Center GmbH
Europaplatz 3
69115 Heidelberg, Germany

The manufacturer's authorised representative in the EU is Springer Nature Customer Service Centre GmbH, Europaplatz 3, 69115 Heidelberg, Germany. If you have any concerns regarding our products, please contact ProductSafety@springernature.com

Printed and bound by CPI Group (UK) Ltd, Croydon, CR0 4YY
30/09/2026
02225319-0003